The BRAIN

The BRAIN

Gerald M. Edelman
Jean-Pierre Changeux
editors

Routledge
Taylor & Francis Group

LONDON AND NEW YORK

First published 2001 by Transaction Publishers

Published 2017 by Routledge
2 Park Square, Milton Park, Abingdon, Oxon OX14 4RN
711 Third Avenue, New York, NY 10017

Routledge is an imprint of the Taylor and Francis Group, an informa business

Library of Congress Catalog Number: 00-044708

Library of Congress Cataloging-in-Publication Data

The brain / Gerald M. Edelman and Jean-Pierre Changeux, editors.
      p.cm.
    "Augmented version of an issue of Daedalus, spring, 1998"—T.p. verso.
    ISBN: 0-7658-0717-3 (pbk. : alk. paper)
      1. Brain. I. Edelman, Gerald M. II. Changeux, Jean-Pierre. III. Daedalus
(Boston, Mass.)

QP376 . B6954 2000
612.8'2—dc21

                                    00-044708

ISBN 13: 978-0-7658-0717-5 (pbk)

Contents

Preface

THE HISTORY OF MODERN SCIENCE has seen a number of conceptual revolutions in many fields. Within the last hundred years, for example, evolution, genetics, physics, geology, and cosmology have undergone vast changes in their conceptual structures. Now, at the millennium, another revolution appears likely in the field of brain science. When that revolution is complete, many longstanding psychological problems and epistemological puzzles are likely to be resolved.

Perhaps the most fundamental of these is the mind-body problem—how the workings of the brain can give rise to perception, memory, feelings, and consciousness. For a long time, problems in these areas were considered to be unsolvable or even to be outside the proper domain of scientific exploration. Indeed, as A. N. Whitehead first noted, since the very beginning of Western science, the mind was removed from nature. Galileo, quite properly, felt it was not necessary to consider the mind in pursuing his physics. And with his dualism, Descartes quite explicitly considered the mind to be a thinking substance, not an extended substance examinable by physics. By the end of the nineteenth century, however, the rise of experimental physiology and psychology made it clear that the mind had to be put back into nature by scientific means. This remains a formidable task and the path to its accomplishment has not been a smooth one. In the present century, we have seen, for example, the rise and fall of introspectionism, of behaviorism, and, more recently, of the computer model of the mind. But now the task of understanding the relationship between the mind and the brain appears achievable.

What is different now is the remarkable explosion of knowledge in neuroscience over the last several decades. A confluence of multiple disciplines and a still unabated swell of interest are almost certainly major contributors to this remarkable growth. The implications for matters of human concern are obvious: Neurological disease, psychological exploration, psychiatry, the human sciences, and even art are all likely to be strongly influenced by the findings of neuroscience.

For this reason, it seemed timely in early 1998 that an issue of *Dædalus*, the Journal of the American Academy of Arts and Sciences, be dedicated to an account of progress in brain research. The goal was to give the non-specialist reader a vision of the history, theoretical underpinnings, clinical promise, and human implications of neuroscientific research. At the same time, a number of contributors, well aware of the possibility of a scientific revolution in the making, provided a view of the future. The present book is based on all of their efforts, and it aims to reach an even wider audience than the *Dædalus* issue.

Modern neuroscience began in connection with medicine, and the contemporary neurology that grew out of this connection is a superbly precise diagnostic engine. But, largely because of the fact that central nervous system neurons do not regenerate, the formation in many diseases of what amounts to scar tissue after cell death prevents the formation of new connections. Modern advances in early diagnosis by techniques of brain imaging and molecular biological characterization of basic cellular processes of the neuron promise to change this picture. Moreover, transplantation methods and implantation of electrodes by modern neurosurgery show great promise in treating diseases such as Parkinsonism. For these reasons, the somewhat gloomy forecasts of a decade ago have been replaced by guarded optimism. Within the next decade, we can expect the same kind of happy transformation to occur in neurology as was experienced in cardiology after World War II.

As significant as this is, the resolution by neuroscientific research of some of the deep problems of psychology will be even more significant. It has now become clear, for example, that a major mechanism underlying learning and memory is a biochemical change in the strengths of neural connections in the brain known as synapses. The intricacy of these structures is stunning and their numbers are hyperastronomical—there are approximately 1 million billion synapses in the cerebral cortex alone. Through studies of synaptic changes, the likelihood of understanding memory and even improving it has recently been enhanced. The mechanisms by which we perceive events and objects and the means by which we control movement are also becoming better understood.

These advances stem in part from new technologies ranging from molecular biological assays, through the use of multiple recording electrodes, to imaging methods applied to the brains of living subjects doing various tasks. It is now possible, for example, to visualize the neural correlates of conscious states. But perhaps more important than any single technical application is the ability to observe a living and behaving animal or human by means of a multilevel approach. A robust theoretical framework that goes well beyond facile comparisons of the brain to a digital computer is now available to make sense of the enormous amounts of data that will emerge. Although much of modern biology is pursued without strong dependence on theory, if brain science is to deal with the very bases of our sentience, it must not take such a stance. There is now an increasing interest in consolidating intermediate range theories involving perception, memory, and consciousness into a global theory of brain function. Several essays in this volume address this important issue.

What are the unmet challenges to this vigorous and fruitful program of research? Above all, we need to resolve some outstanding mysteries. One is consciousness itself—a problem whose solution, William James suggested, would mark the greatest scientific achievement of all times. A second problem is a related one—why do we sleep? Much progress has been made in this arena but the actual function of sleep remains a mystery. Another area of deep social significance relates to drug abuse. How can we meet the challenge to human integrity that is posed by addiction? And, of course, there remains the vast field of mental disease that is connected in one way or another to all of these extraordinary problems.

A key area to be addressed by neuroscience concerns the expressive or emotive aspects of brain function. These aspects are seen and experienced as feelings, and, in language, as the communication of feelings. Analogies to physics and computationally based brain theories either do not address the origin of feelings or fail abysmally to account for them. This stands as a major criticism of those aspects of cognitive psychology that are based on the computer model. There is a suggestion that the so-called selectional brain theories described in this volume may be able to address this issue. Such theories depend on the workings of evolved value systems of the brain and their task is to connect these basic neural systems to emotion and meaning. If they succeed, the insights that emerge will greatly affect approaches to improving the development and education of children.

The essays in this book represent a broad sample of current work in neuroscience. They provide a variety of assessments of the significance of neuroscience for the matters of human concern that we have touched on here. While there will undoubtedly be debate about whether the neuroscientific revolution has actually begun, the distinguished contributors to this volume do not doubt its ultimate significance. In any event, the contents of this should allow the reader to make his own judgments on the matter and, at the very least, provide a vision of a fascinating scientific journey.

Gerald M. Edelman and Jean-Pierre Changeux

Vernon B. Mountcastle

Brain Science at the Century's Ebb

I OPEN THIS BOOK on the brain with a brief philosophical comment. The reason is clear: the half-century's accumulation of knowledge of brain function has brought us face to face with the question of what it means to be human. We make no pretension that solutions are at hand, but assert that *what makes man human is his brain*. His humanity includes those aspects of behavior traditionally classed as mental. What is of signal importance is that some aspects of the inner life of man are now open to scientific inquiry. Few neuroscientists now take a non-naturalist position, and still fewer hold to a principled agnosticism on the mind-brain question. The vast majority believe in physical realism and in the general idea that no nonphysical agent in the universe controls or is controlled by brains. Things mental, indeed minds, are emergent properties of brains.[1] Those emergences are not regarded as irreducible but are produced by principles that control the interactions between lower level events—principles we do not yet understand.

How this is so is a formidable experimental task in neuroscience, for we face a large explanatory gap between the processes of brain and the processes of mind they produce. Still, the gap is beginning to narrow. What is most in need of an explanation—or, in the beginning, even the outline of what an explanation might look like—is consciousness itself. Perhaps we should seek an explanation for one of the simpler of the several levels of consciousness that are still nebulous, for there appears

Vernon Mountcastle is Professor Emeritus of Neuroscience at the Zanvyl Krieger Mind/Brain Institute at Johns Hopkins University.

1

to be a consensus among cognitive scientists that consciousness is not a unitary function or process.[2] My own opinion is that the word "consciousness" designates a supraordinate class of phenomena that includes several different states having a major general property in common. We are conscious of self, aware of sensory stimuli, aware when we generate action, and aware of our internal operations (like thinking). None of these forms of awareness are likely to be identical, yet it is the general property of "being aware" that is common and that allows us to classify them together. One candidate for study at the level of brain mechanism is primary sensory-perceptual awareness, that is to say, "phenomenal" or primary consciousness: How are we aware of objects and events in the world around us, of our own intentions to act upon the world, of our own internal operations and sets? Some believe that conscious awareness is qualitatively different from those things humans can comprehend and is thus forever "cognitively closed" to us.[3] I know of no neuroscientist who believes this. Contrarily, in spite of much confusion and a plethora of lengthy volumes, confidence is strong that through persistent study of the intermediate brain operations that lead from external events to our vivid conscious awareness of them, and from the conscious will to act to extrinsic action, we shall gradually close in on the target—the neurological mechanism of conscious awareness.

Isaiah Berlin, in an essay published just before his death, wrote: "The ideal of all natural science is a system of propositions so general, so clear, so comprehensive, connected with each other by logical links so unambiguous and direct that the result resembles as closely as possible a deductive system, where one can travel along wholly reliable routes from any point on the system to any other."[4] No one imagines that neuroscience yet constitutes such a system, but the advances of the last half-century in methods and concepts, and the accumulation of facts about the brain and how it controls behavior, have brought many general principles onto the field of play. One can perceive, even if dimly, such a system of propositions—"so general, so clear"—that characterize the present state of brain science.

It is possible to identify at least eighteen disciplines whose concepts, methods, and practitioners tend towards union in the current panorama of the brain sciences. They naturally fall into groups of two or more areas that appear more closely related to each other than to those of other sets: neuroanatomy, neurophysiology, and biophysics; cellular and molecular neurobiology, genetic neurobiology, and neurochemistry; evolutionary and developmental neurobiology; experimental psychology and psychophysics; neuropsychology, the clinical neurological sciences, and neuropharmacology; classical cognitive science, cognitive neuroscience, and computational neuroscience; and some areas of epistemology and philosophy. The boundaries between some of these disciplines are to a degree artificial, as the general concepts, methods, and indeed the investigators themselves move effortlessly between disciplines. Yet this merging of previously isolated disciplines into the many unified components of modern neuroscience is of profound significance for its practice and its future.

I aim in this essay to highlight some of the major contributions to knowledge of brain function made during the last half-century, especially those that reveal some general propositions.

THE NERVE IMPULSE

Brains contain two major classes of cells. *Neurons* are the active signaling elements of the brain. *Glial* cells, which equal or exceed neurons in numbers, do not transmit signals but serve a number of other functions. Some act as supporting elements of the brain; another class synthesizes the myelin sheaths that wrap most of the larger nerve axons of the brain in successive layers; still others act as scavengers, removing the cellular debris that follows injury or the ubiquitous cell death within the brain, or buffering the increases in extracellular potassium concentration that accompanies intense signalling activity in neurons (see figure 1, plate 2).

While many different classes of neurons exist in the brain, they all conform to a general pattern. The cell body contains the nucleus and a variety of intracellular organelles involved in the synthetic functions of the cell. Each cell body emits a system of profusely branching structures called dendrites, which receive the majority of the synaptic contacts—the input signaling pathways—from other neurons. Each neuron also emits a single axon that extends for distances that vary from a few millimeters in the forebrain to perhaps 150 cm for the long fiber tracts that reach from the motor areas of the cerebral cortex to the lower spinal cord. The axon is the output channel of the neuron, carrying signals that move along it and invade all of its branches that terminate at other neurons (its synaptic targets) to excite or inhibit them.

Explaining nerve impulse conduction, or how axons conduct signals from one end to the other over considerable distances, has been the object of intensive study for more than a century, since Helmholtz discovered that it has a finite velocity. Understanding this process and the mechanisms of signaling between nerve cells, called synaptic transmission, are undoubtedly the most important general principles of neuroscience to be established in the second half of the twentieth century. They are fundamental to understanding the many other brain functions described in this and later essays in this volume. For generality and clarity they meet Berlin's criteria.

It is important to explain these principles briefly. Neurons, like cells in other systems of the body, have cell membranes consisting of a double layer of lipids sandwiching a layer of protein. The cell membranes of neurons are only 50 to 70 angstroms in thickness, yet they are relatively impermeable to the movement across them of small charged ions, particularly sodium ($Na+$), potassium ($K+$), calcium ($Ca++$), and chloride ($Cl-$). The double bilayer contains a number of protein molecules that extend across the membrane from one side to the other. Some of these proteins contain a *channel* that is open or closed depending on the instantaneous molecular configuration of the protein, which is itself under the control of the membrane potential. Each of these voltage-sensitive channels is selective for one, or in some cases several, of the small ions listed above; it is now known that many other channels with different functions and different ionic selectivities also exist in neuronal membranes. When a portion of a nerve cell is stimulated, there results a decrease in the membrane potential of the nerve fiber, and the virtually instantaneous opening of the $Na+$ and $K+$ channels. Opening its channel allows $Na+$ to flow rapidly into the cell, producing a decrease in the membrane potential to zero and to its reversal. This allows K+ to move from inside to outside, which repolarizes the axon. This is the "depolarization-repolarization" sequence of the nerve impulse. It is conducted along the axon because each successive small segment of the axonal membrane is depolarized by the movement of charge from its surface into the depolarized region. The velocity of the nerve impulse is constant along the axon, determined by axon diameter. Organisms with nervous systems thus depend for life itself upon the slight temporal stagger between the time courses of the changes in membrane permeability to Na+ and K+. This principle derives mainly from the discoveries of Hodgkin and Huxley, described in their five papers of 1952.[5] It evolved from a long history of experiments on the nerve impulse—from the time that Helmholtz proved it to have a finite conduction velocity, through the work of Julius Bernstein (around 1900), to that of many others, particularly Kenneth Cole.[6] It was the combination of several classes of facts, methods, and concepts from different sources that led to these major discoveries.

ON SYNAPTIC TRANSMISSION

The human brain has a volume of about 1,370 ml; its neocortex contains about 28 billion neurons connected with each other by large numbers (on the order of one to ten trillion) of contacts called synapses.[7] The brain is a system of complexly interconnected neurons organized into local microcircuits, which are then linked into distributed systems. A central dogma of brain science is the neuron theory, which says that the linkages between nerve cells are in contiguity, not continuity, and that the business of the brain is executed by the transmission of signals between these contacting elements, a special process called synaptic transmission.[8]

The change from phenomenological descriptions of synaptic transmission to explanations at the level of cellular events accelerated in the early 1950s, has continued at a dazzling pace since then, and—together with the ionic hypothesis (illustrated in figure 1)—has created the most important generalizing propositions in neuroscience. This body of knowledge forms the basis of concepts ranging from the explanation of spinal reflex action to the brain mechanisms of memory and learning. They have spawned the discipline of neuropharmacology, which has contributed to the new biological psychiatry. How did this wonder come about?

First, the intensive study of the problem followed the era of Claude Bernard in the last century. This work culminated in the discoveries of the 1930s and 1940s, largely by pharmacologists, that synaptic transmission depends upon the release of a transmitter substance from active nerve endings, the union of the transmitter with a special "receptive substance" on the membrane of the postsynaptic cell, and the activation of that cell by the combination.[9] The transmitter acetylcholine was identified at the neuromuscular junction and at some autonomic synapses, while the transmitter norepinephrine was present at others. The molecular details of these synaptic actions were then unknown, but the broad outline of the sequence of synaptic events has stood the test of a host of experiments since that time.

The critical experiments were done almost simultaneously in the laboratories of Bernard Katz and John Eccles, using different experimental models. In the first the methods described above were used to study synaptic transmission at the junction between frog nerve and muscle. The results obtained established the sequence of events linking nerve action potentials to muscular contraction. Briefly, transmitter molecules are synthesized in neuronal cell bodies, transported along axon terminals, and stored in small vesicles. Invasion of the terminal by a nerve impulse initiates a massive influx of calcium into the terminal, resulting in the release of vesicles and their contents into the intracellular cleft between the synapting cells. The union of the released transmitter molecules with receptor proteins exposed on the surface of the postsynaptic cell in turn elicits changes in ionic conductances across the membrane of that cell. The generalizing principles of chemical transmission are almost universally applicable to chemically operated synapses, and have been challenged only in details in the nearly fifty years since their description, although many variations on this classical model have been proposed.[10]

THE GENERALIZING PRINCIPLE OF SYNAPTIC PLASTICITY

Since it was first described half a century ago, the field of synaptic transmission has been expanded and sustained by the revolution in molecular neurobiology.[11] Of many new discoveries, perhaps the most important is the discovery of synaptic plasticity and descriptions of its candidate mechanisms. Synaptic plasticity has emerged as an important generalizing proposition in neuroscience; in sum, it holds that the effectiveness of transmission at neuronal synapses and indeed the initial formation of synapses during development depends upon the level of synchronous activity in the presynaptic and postsynaptic cells. Synaptic plasticity is an important factor in the ontogenetic development of neural circuits in the cerebral cortex. After sets of axons afferent to the cortex are guided in the selection of pathway and general target, they face the problems of making connections to the correct spatial location on the surfaces of target neurons and the stabilization of those connections. The stability and later refinement of the initial synaptic contacts relies in part upon activity-dependent competition between axons aimed at common postsynaptic target neurons. Initial synaptic contacts are strengthened when there is phase-related concurrent activity in pre- and postsynaptic elements, and weakened when there is not. Previously thought to be confined to a critical period early in brain development, these properties of synaptic plasticity and modifiability are now believed to be virtually universal properties of brains, young or old.[12] It seems likely that there are ranges of brain structure and function that remain unchanged throughout adult life; they set the individual identity of the organism.

The inference can be drawn that, at the level of synaptic microstructure, *no two brains are identical and any single brain is continually changed by its experience.* A major theme in current research programs in molecular neurobiology is the discovery of the mechanisms of synaptic plasticity, which is regarded as the cellular basis of such global brain functions as learning and memory, motor learning, the cortical mechanisms in perception, and so forth; it is a fundamental premise in several general theories of brain function, including that of neuronal group selection described in a later essay in this issue of

Dædalus.[13] The action of the afferent modulatory systems of brain-stem origin in controlling long-term changes in excitability of the forebrain, such as those of the sleep-waking cycle, are described in another essay in this issue. It is important to emphasize that the modulatory afferents can change the processing operations in both cortical microcircuits and in distributed systems.

ON ONTOGENESIS

The three-dimensional structure of the cerebral cortex into horizontal layers and interdigitated vertical columns running across those layers is generated from a two-dimensional array of progenitor cells in the epithelium lining the embryonic cerebral ventricles. The progression of events from cell birth to the specification of cortical areas is determined by the temporal order in which certain genes are expressed, often in combinations that change with time. A current research goal is to determine the sequential and interactive roles of intrinsic ("cell-autonomous") and extrinsic ("environmental") factors that control developmental events. Cortical neurons migrate from their loci of generation to their final positions in the developing cortex, moving along palisades of guide fibers consisting of modified glial cells that form a temporary scaffold.[14] This provides a match in the x-y dimension of the germinal epithelium with that in the overlying, developing cerebral cortex. The migration of immature neurons begins shortly after the first asymmetric cell division of the progenitor cells in the germinal epithelium. Once they reach their destinations, the waves of migrating neurons form the cortical plate, which consists of narrow chains of neurons, called ontogenetic units, oriented normally to the cortical surface.

After migration, neurons differentiate further, project outgoing and receive incoming nerve fibers, and make local connections that form the network basis of the vertically oriented columnar networks of the mature cortex. They generate molecular transmitter mechanisms for synaptic signaling and synthesize receptor molecules that allow a response to transmitter signals reaching them from other neurons (see the explanation accompanying figure 1). Ontogenetic units mature into minicolumns;

these are grouped into columns, and columns into areas. A cortical area contains a large number of columns grouped together, possesses a distinctive cytoarchitecture, and receives a defined innervation from a central structure called the thalamus; the neurons located in the thalamus have a distinctive set of functional properties. The initial specification is loose, and ongoing dynamic activity in the afferent systems makes more precise the boundaries and sizes of cortical areas, as well as the development of specialization in these areas.

The rapid accumulation of knowledge about the ontogenesis of the mammalian neocortex ranges from the molecular mechanisms in cell differentiation, cell-type specification, and neuronal migration to the interaction of genetic and environmental influences in defining the connectivity within and between cortical microcircuits and the specification of architectural fields. It now appears possible to define and study in nonhuman primates how genetic mutations and environmental insults produce devastating developmental abnormalities in human brains.

ON PHYLOGENESIS

All cerebral cortices of extant mammals consist of six layers and display a modular organization, and while the constituent neuronal cell types differ significantly in detail they all use widely overlapping sets of synaptic transmitter agents and receptors. Living mammals of several radiations elaborate large cortices by enlarging primary sensory and motor areas and elaborating new cortical areas; identified areas vary from ten to twenty in small mammals to perhaps one hundred in humans.

Perhaps there is no more startling event in evolutionary biology than the enlargement of the brain in hominids. *Hominidae* are thought to have emerged from the stem *Hominoidea* during the Miocene-Pliocene transition period, perhaps five to seven million years ago, by a split from a common ancestor that produced australopithecines and pongids (apes and chimpanzees). A dramatic episode in our history followed: the early australopithecines assumed the upright stance and bipedal locomotion. Perhaps they were not the first primates to do so and perhaps they did so intermittently, but they initiated the chain of events leading to *Homo sapiens*. They did so with brains no larger than those of modern pongids, but their new way of movement must have required adaptations in the dynamic operating characteristics of central motor-control systems that might not be evidenced by changes in brain size or external morphology.

A tripling of brain size from 400 to 1,300 grams marks hominid brain evolution, which is associated with increased manipulative skill in throwing with arm and hand, the acquisition of a stone-tool culture, family life, and, perhaps only much later, a spoken language and the development of the culture and technology of modern civilization. Some changes in brain structure accompanied its enlargement: expansion of the proportional sizes of the posterior parietal and frontal cortices, the appearance of a motor-speech area, and increased bilateral asymmetry. Based on evidence from fossil endocasts, the brain of *Homo sapiens* has changed not a whit in size, form, or external morphology since the species first appeared less than a hundred thousand years ago. It is difficult to believe that man's transition from a cave-dwelling scavenger to the creator of modern civilization has been accomplished without some important change in the operating characteristics of his brain. Such an evolutionary change could only occur within the parameters of a brain structure already in place, and the modifications in the mode of dynamic processing might not be revealed by changes in size or external morphology. This emphasizes a major research objective in the neurosciences—to study the human brain directly with all the methods available while still forging ahead with the development of badly needed new methods. It is obvious that the human brain is not a scaled-up model of the 80-gram brain of the macaque monkey, much less that of a rodent. A question of interest is whether the long-standing dogma of neuroscience that neuron cell types and their biochemical mechanisms are strongly conserved in mammalian evolution is universally true, especially if it applies to the human brain. Some neuron types in the human brain do differ from those of other mammals, particularly in their biochemical mechanisms. Moreover, it appears that the patterns of connectivity between cortical areas and between the cortex and other brain structures may not be simple, expanded versions of those in the monkey brain. The hypothesis to be explored is whether the greatly expanded cognitive capacities of the human brain, and the development of a "theory of mind," depend upon something more than increases in brain size and neuronal number.[15] It seems safe to predict that properties unique to the human brain, particularly in dynamic patterns of operation, will be revealed by future research.

ON STRUCTURE

Prominent among the discoveries that led to the emergence of brain research as a scientific discipline are the findings that the mammalian neocortex is a laminated structure and that the distribution of neuron types and their packing densities differ between cortical areas (a form of histological description called cytoarchitecture). The results were detailed parcellations and eventually maps of the cortex of many mammals, including humans. The general position was reached that the transition from one cytoarchitectural field with a certain set of neuronal distributions to an adjacent field with different ones might be abrupt, that the changes occurred almost simultaneously in all the cellular layers, and that the transitions could be marked with a line. It is remarkable that these two principles remain virtually intact more than a century after their discovery, even though the criteria for defining cortical areas now include their patterns of extrinsic connections and the functional properties of their intrinsic neurons.[16]

Few would have guessed at midcentury that the study of structure would be among the most successful of research efforts in neuroscience in the decades that followed. Yet this is indeed the case, for the gradual understanding of the morphology of cortical neurons, the microcircuits in which they are embedded, the connectivity between brain parts (especially between areas of the neocortex), and how these systems are organized in laminar, columnar, and distributed-system arrangements has produced a model of the brain that our colleagues of two generations ago would scarcely recognize.

The intracellular injection of molecular markers has established the specificity of neuronal cell types in the cerebral cortex and other regions of the brain, allowing direct correlation of their functional, biochemical, and morphological properties. An expanding area in neurobiology is the use of biochemical and immunohistochemical methods to identify molecules within cells and on their surfaces that are differentially specific for different neuron types, and in some cases for cortical areas and systems as well. These results point to a future chemical architecture of the brain.

A new fact about large-scale brain anatomy is that the number of connections that link brain structures, and particularly those between areas of the cerebral cortex, are perhaps on an order of magnitude greater than had been supposed hitherto. For example, more than eight hundred pathways have been identified connecting the seventy-plus areas in one hemisphere of the macaque monkey brain. This increment in knowledge has been generated by use of methods based on the discovery that molecules and small intracellular organelles (like vesicles) are transported in each of the two directions between the neuronal cell body and its axon terminals in a process called axoplasmic flow.[17] These connections are so dense and widespread, though specific in terms of sources and targets, as to support a shift to the view of the brain as a collection of many interacting distributed systems and a modification of the traditional view of staged hierarchical processing.

Intuitive images of this complexly interconnected system have remained informal and speculative. Recently a number of methods have been used in attempts to analyze it in quantitative terms; among these are hierarchical analysis and both metric and nonmetric multidimensional scaling.[18] The first is based on the fact that the neuronal projections linking cortical areas have different laminar source-target relations and are classed as ascending, descending, or lateral. These rules are then used to arrange cortical areas in an ascending ladder, beginning with the first cortical representational areas of sensory systems and progressing into the "higher-order," homotypical (association) cortex. This analysis has yielded reasonable arrangements of areas at the early input stages of sensory systems, but the placement of areas at one or another level in the ladder is to a degree a matter of choice, for a number of arrangements appear to fit the rules equally well.

Nonmetric multidimensional scaling exploits the fact that qualitative data in sufficient numbers can yield quantitative solutions. In the present case it has been used to produce a metric structure that reveals the relations between cortical areas in a topological space—a connectional or processing structure independent of cortical geography. The result is an image of the cortex set in an equilibrated tension in which the distance between cortical areas is determined by the inverse of the density and number of the connections between areas rather than actual geography. The results confirm the hierarchical nature of areal arrangement at the entry levels of sensory systems into the neocortex and the following change to a distributed format in the large areas of cortex beyond the primary sensory and motor areas.

ON PERCEPTION

An ancient problem in natural philosophy and a major research program in present-day neuroscience is to determine the relations between the material order of the world around us and the sensory-perceptual order of our experience. The experimental objective is to discover the brain mechanisms of the intervening neural transformations. The primary features of the stimuli that impinge upon us—the "distal" stimuli of heat, force, light, sound, and chemical substances—are selectively transduced at the peripheral ends of sensory nerve fibers. Some groups of fibers respond more selectively at lower thresholds than do others to different forms of impinging energy. This tuning, sometimes called feature detection, is accomplished by specific transducer mechanisms either directly in the nerve endings themselves or in complex sensory organs in which they terminate. Our perceptual experiences are thus mediate, once-removed, slightly delayed in time, abstracted images determined by the transducing properties of the receptors and the processing properties of the central neural networks they engage.

Nerve impulses are projected into and relayed through the sensory pathways of the nervous system, susceptible at every synaptic level to transformations imposed by the microstructure of the relevant neuronal populations and by the regulating influence of neural systems of central origin. The major sensory systems project into the primary sensory areas of the cerebral cortex in topographical mapping of the relevant sheets of sensory receptors, and thus of the portion of the sensory world each subtends. These are called representations, a word used in neuroscience to convey the idea of a central neural reflection of some particular aspect of the sensory world, frequently shown in the topographic mapping of sensory sheets at successive levels of the central nervous system.

Cortical maps allow sustained representations of sensory and motor events in an accurate topology, in the midst of ever-changing and slightly shifting population actions. They maintain the near-neighbor relations that persist through the spread of activity in distributed systems and the death of neurons, and thus allow sustained representations. They also offer the possibility of plastic changes in the mapping produced by changes in behavior, or in acutely or chronically altered input.

Some signals lead more or less directly to motor output, executing the spatial coordinate transformations linking sensory and motor dimensions. It is surmised that the input representations are combined with stored records of past experience, leading in the one case to perception and in the other to neural patterns that upon discharge drive movements; in many cases the two are inextricably linked. Central neural representations vary along a continuum from those isomorphic to physical stimuli or movement patterns to those that are more abstracted.

The most productive method used in studies of the brain mechanisms in perception is the combination of psychophysical measures of the performances of humans and nonhuman primates as they work in perceptual tasks, with simultaneous recording of signs of brain activity evoked by the task stimuli and the subject's responses. The aim is to identify causal relations between behavioral and brain events, and to determine which of the latter can be shown on other grounds to be necessary and sufficient for the former. The methods used vary from measurements of changes in blood flow in local regions of the brains of human subjects, to the recording of electrical activity in humans by electro- and magnetoencephalography, to the use of acutely inserted or chronically implanted multiple microelectrodes in the brains of nonhuman primates.

The signals of different stimulus attributes are projected over afferent systems to, and processed within, separated but connected cortical areas that are specialized for modes of processing of one or more attribute, rather than for functional localization. No "higher-order" target zone has been discovered to which these separated processes project, and in which an integrated neural image of a perception might be constructed. One hypothesis presently under intensive study is that such a neural image is embedded in the dynamic activity of the distributed system itself. This hypothesis has raised the binding problem: How are the neural activities within and between nodes of a distributed system bound together and recognized as a coherent representational state of the perception? Consider, for example, the binding problem between visual, proprioceptive, and tactile inputs guiding motor activity as a projecting hand begins to conform in midair to the spatial form of its target, and completes the task guided by tactile input after target acquisition. The solution proposed is that binding is enabled by a transient synchronization of the neuronal activity in the several local regions of the system, evidenced by synchronization of neuronal impulse discharge. The binding hypothesis is limited, however, for it offers no solution to the perennial problem: what next? That is, what neural mechanism can be imagined to recognize the presence or absence of synchronization at a particular frequency, and to identify the pattern as that evoked by a particular external event?

Synchronized activity has been observed between two locations in the same cortical area, and between two different areas in the same hemisphere and in the two hemispheres. The activities are synchronized only if linked by stimulus features common to the processing specializations of the areas involved.[19] While these observations support the binding hypothesis, they are not universal, for they have not been seen in the experiments of other investigators. A set of experiments is needed in which neural events are recorded in several nodes of a distributed cortical system of a working monkey and compared on a trial-by-trial basis with the behavioral outcomes.

Investigations of many sorts using the combined experimental approach described above yield increasingly detailed descriptions of cortical neuronal events during perceptual operations. The population actions are of great interest, for there is no single cell capable of representing, for example, a "smiling grandmother," the so-called grandmother cell. Presently, studies appear constrained to the broad range of preconscious processes; indeed, if the neural mechanisms of conscious awareness have ever been recorded, they have not been recognized as such. The final products of these neuronal operations flow seamlessly across the threshold to conscious perception, fait accompli. The intermediate mechanisms at the level of microcircuits and distributed systems are subjects of much present-day research in neuroscience, sustained by the belief that until we have a better understanding of these intermediate processes those of conscious awareness will remain obscure, absent some unpredictable discovery. General principles linking patterns of neuronal activity to perceptual events are badly needed.

WHAT ARE THE BASIC FUNCTIONS OF THE CEREBRAL CORTEX?

The direct answer to this question is that no one knows. From what has gone before the reader might infer that a large fund of knowledge exists concerning the function of the cerebral cortex. This is not so, for the knowledge available is phenomenological, with few explanations at the level of mechanism. What exists in profundity is knowledge of the geography of the brain and the connectivity within it—answers of twentieth-century neuroscience to *where* questions. Such answers reveal little about the operating characteristics of brains. It is knowledge of the mechanisms of the ongoing dynamic activity of large neuronal populations in brain systems that we need for an understanding of brain function.

The nature of the dynamic operations in local cortical microcircuits and in the distributed systems of the cortex are two of the major unsolved problems in neuroscience. Many functions described at the phenomenological level have been attributed to the cerebral cortex, among them thresholding, amplification, the derivative function, feature convergence and new feature construction, distribution, coincidence detection, synchronization, long-term storage and retrieval, and so on. Yet no one of these is fully understood at the level of neuronal-circuit operation. The most intensively studied of the initial local-circuit operations is that at the entry level of the primary visual cortex. Binocularity may be explained by simple convergence of inputs from the two eyes, but the question has not been settled as to whether the response to stimulus orientation is produced by a similar projection-and-convergence mechanism, as some believe, or whether such a convergence is combined with a microcircuit operation somewhere between that enhances this feature.

The second problem is set on a larger scale: What are the neuronal operations within the distributed systems of the cortex? Here the phenomena of feature convergence and new feature construction reach their epitome in the distributed systems of the parietal, temporal, and frontal lobes. For parietal neurons, combinations of sensory and motor features are integrated with the neural mechanisms of central-drive states; the full set of features is displayed only when the behavioral task in which the monkey subject is engaged has meaning for him.[20] The posterior parietal cortex receives converging, relayed projections from several visual areas, as well as from the retinocollicular system and many intense interhemispheric projections. No single one of these sources has neurons that display the full-feature profile of parietal visual neurons: large, bilateral receptive fields; sensitivity to motion and its direction with radial orientation of directionalities; and similar characteristics. While these functional properties of the parietal distributed system are well-known phenomena, the local cortical operations producing them are unknown.

These two questions deal with actions in neuronal populations that vary in number from about one hundred neurons in a cortical minicolumn to the millions and billions in distributed systems. They have one property in common: the dynamic properties of the actions of integrated neuron populations cannot be deduced from knowledge of the action of the single neurons within them. Instead, the population properties are emergents. The neuronal circuits of the cortex are embedded in the matrix of the brain, and no method has been devised for the observation of local neuronal circuit action during a transcortical operation. Consequently, the functions of cortical areas are still defined in terms of the external connections of those areas, as they have been since the golden era of clinical neurology. Yet it is obvious that there is nothing intrinsically motor about the motor cortex, nor sensory about the sensory. The similarity of lamination pattern, neuronal cell types, and intrinsic connectivities has led some to suggest that the intrinsic operations in different cortical areas are similar, and that the different functions attributed to different cortical areas are largely due to differences in their extrinsic connectivity.

A method used for study of the intrinsic mechanisms of microcircuits is to record intracellularly from neurons in a cortical-slice preparation present in a dish, or in the intact brain, in order to observe (so far as is possible under these restricted conditions) the functional properties of the neuron from which recordings are made. The procedure is to inject into the subject neuron a molecular marker that outlines all its extended branches—some of which extend for millimeters—then recover the cell and reconstruct it anatomically.[21] Double intracellular stimulation and recording in cortical-slice preparations identifies precisely local intracortical synaptic connections. This research is providing detailed templates of cortical microcircuits. To study the dynamic operations in such microcircuits we need methods for recording from a substantial number of the hundred or so neurons in a cortical minicolumn of a waking monkey as he executes a behavioral task relevant for the cortical area under study. For study of the dynamic operations in extended distributed systems, imagine one thousand microelectrodes chronically implanted to cover the several nodes of the parietofrontal system as a monkey reaches to a target he has selected in a perceptual decision. Such an experiment will mark the transition of cortical neurophysiology from little to big science, with experiments executed and the results analyzed by a team of scientists who bring to the common effort a broad range of skills. Such an experiment is of transcendental importance, for absent new technical developments no other experimental approach is obvious.

THE HUMAN BRAIN SCIENCES

We have witnessed in recent decades the rapid coalescence of a number of disciplines around the central theme of study of the human brain in health and disease, and of dynamic brain activity in waking humans as they execute cognitive, perceptual, or motor behaviors. This has converted the clinical neurological specialties to scientifically based disciplines and paved the way for objective study of how human brains generate and control human behavior, including that at the most distinctively human levels.

Perhaps the most human of sciences is that concerned with the nature of knowledge and how we come to acquire and use it—a theme at the heart of the Western philosophical tradition. Cognitive science is thus both ancient and new, developed in its modern form in the last half-century. Its central theme is the concept of mental representations, and the dominant model, the electronic computer. This enterprise has been successful in generating computational analyses and models of how we think, speak, remember, imagine, attend, solve problems, and like phenomena. Cognitive scientists bring to these problems facts and concepts from several disciplines, including cognitive and experimental psychology, linguistics, computational theory, anthropology, artificial intelligence, and the philosophy of mind. They explore cognitive operations at three successive levels: implementation; algorithmic operations over symbolic representations, according to rules; and formal computation. The symbol/rule approach led to the proposition called "multiple realizations," which holds that a given cognitive operation could be executed by any of a variety of computing machines, not excluding but not limited to biological ones, and that how cognitive operations are executed in human brains is not a question for cognitive science. One authority expressed the strong view that "a central idea of modern cognitive science is that the human cognitive system can be understood as if it were a giant computer engaged in a complex computation."[22] In spite of this metaphor, it is clear that cognitive science overlaps with and is an important part of the human brain sciences, and will continue to make important contributions to them. One reaction to the computational metaphor was strong and swift, leading to the development of the related but different cognitive neuroscience, which I describe in a later section.

Imaging the Brain by Recording Its Electrical Activity

A major change in studies of the human brain has been the perfection of methods for recording signs of global brain activity in waking human subjects—signs of brain activity that correlate to brain states at the phenomenal level. The electrical oscillations ("brain waves") recorded from the surfaces of the head or brain in mammals are changes in electrical potential called the electroencephalogram (EEG). The EEG records preferentially the cranial potential landscape generated by currents flowing radially (normal to the cortical surface); the magnetoencephalogram (MEG) records those flowing tangentially. The prevailing orientation of the pyramidal cells, whose actions generate the recorded extrinsic current flows, suggests that the EEG is generated largely by activity in surface gyri of the cerebral cortex, the MEG by that in the banks of cortical sulci. The electrical oscillations vary in frequency over the range 1 to 100 Hertz, and from a few to a few hundred microvolts in amplitude. They persist in characteristic form during the states of normal alertness, drowsiness, sleep, coma, or epileptic episodes, never ceasing short of cerebral catastrophe or impending death. The mechanisms producing the EEG/MEG include in varying proportions afferent input over thalamocortical systems, the intrinsic oscillatory properties of some sets of cortical neurons, and the rhythmic activity in local and distributed cortical networks. In addition to its relay function, the dorsal thalamus also operates as a neuronal oscillator and drives the rhythmic activity of the cerebral cortex.

It was Hans Berger who took up Caton's 1875 discovery that changes in electrical potential are generated at the surface of the brain. In the period stretching from 1901 to 1936 Berger succeeded in recording the human EEG; he also discovered the alpha rhythm (10–13 Hz) of the resting brain and its block by the onset of alertness or cognitive activity. Berger proposed that explanatory correlations be sought between the oscillating wave patterns of the EEG, brain processes, and human behavioral states. He thus described the first combined experiment, setting with a single stroke the long-term research program of electroencephalography.[23] In the sixty years since Berger's time his program has been expanded and intensified by neuroscientists who sought the correlations Berger envisaged. The EEG patterns characteristic of such states as sleep and wakefulness, coma, and anesthesia are now well known. Use of the EEG has revolutionized the physician's care of epileptic patients and is of value in the study of many other pathological brain states; these successes led to the important clinical discipline of electroencephalography.

EEG recording with standard methods has limited utility for specifying the spatial and temporal properties of brain operations during cognitive functions. A major reason is that with classical methods of recording and analysis neither the EEG nor the MEG can be devolved to a unique solution. Any given pattern of surface recorded activity may be produced by more than one distribution of neural activity within the brain. This outlook has been changed by new methods for recording and analysis. The spatial resolution of sources of electrical potential changes in the brain has been reduced to the low mm range by recording with many (in excess of one hundred) electrodes, by registration of recording locations with images of each subject's brain obtained from magnetic resonance images, and by eliminating through analytic methods the blurring effect of volume conduction through scalp and skull. Temporal resolution has been reduced to the low msec level. It appears likely that with these new methods the EEG and MEG will be useful in the field of brain imaging, particularly when combined with imaging methods based on other variables.[24]

Brain Imaging by Recording Changes in Blood Flow

Another set of imaging methods are based on the changes in blood flow and oxygenation that colocalize with increased neuronal activity.[25] The linking mechanisms, and the relations of the spatial and temporal contours of the two, are still uncertain. Nevertheless their empirical association is the basis of a number of successful methods for noninvasive mapping of activity patterns in the waking human brain, especially in the cerebral cortex. The first and most widely used of these is positron emission tomography (PET), with which measures can be made not only of blood flow for imaging purposes, but also of molecular events in brain-energy metabolism and in transmitter-receptor systems. PET requires the inhalation or injection of radioactively labelled tracer molecules, with consequent limitations in repeatability, and is thus not completely noninvasive. The use of PET has produced a basic fund of information about the functional anatomy of the brain and how it changes in different behavior states, particularly when humans execute cognitive tasks.

Nuclear magnetic resonance (NMR) is a method for noninvasive, chemically specific studies of living brain tissue, particularly for many aspects of brain biochemistry (notably of glucose and oxygen metabolic rates), and for monitoring changes in blood flow associated with changes in neuronal activity. Magnetic resonance imaging (MRI) is a method for visualizing living tissue based on the NMR signals from the protons of tissue water. The paradoxical fall in the concentration of the paramagnetic molecule deoxyhemoglobin with increased neuronal activity decreases the interference by that molecule with the NMR signal from water protons. Thus the local water signal increases with increases in neuronal activity. The method produces three-dimensional images of brain anatomy unmatched for clarity and detail, so that functional MRI (fMRI) is rapidly becoming the method of choice for imaging the active brain. Functional MRI is totally noninvasive and can be repeated rapidly in the same subject; moreover, its spatial and temporal resolving power are steadily improving. These methods evolve rapidly, and it seems likely that in combination with brain EEG and MEG we will soon be able to follow in real time and with sharp spatial resolution the changing patterns of cerebral active states. They are described in a later essay in this issue of *Dædalus* by one of the principal architects of the field.

Cognitive Neuroscience

This new field was formed by a union of certain aspects of cognitive science with the neurobiology of the human brain—particularly with methods for imaging its cerebral cortex, described above—with the aim to discover how human brains function when cognating. Cognitive psychologists and neuroscientists to whom this idea is feasible and attractive make experimental designs drawn from cognitive architectures of perception, memory, language, attention, and like activities, and embody them in tasks that human subjects can execute. It is important to emphasize that cognitive neuroscientists are eclectic in their modes of experimental design and methods of execution. However, the coincidence of this new field with the appearance of the new methods for imaging the working human brain has established cognitive neuroscience as an important enterprise in the human brain sciences, one in which mapping studies using imaging methods now dominate the scene. Thus the disciplines of cognitive and imaging neuroscience coalesce. Their aim is to map at high resolution the parts of neural systems, particularly those of the cerebral cortex, that are active as subjects execute cognitive tasks. More recently the combination of methods has been used to add a temporal dimension, to track how active regions shift locations during task execution. The locations of active zones in the cortex are defined by indexing them to unfolded, two-dimensional cortical maps generated from the same brains by fMRI imaging. This combination has revealed a dynamic geography of the brain in action, a functional anatomy of perceptual, motor, and cognitive processes. A very large literature has rapidly accumulated; I summarize here only a few of the important discoveries made.[26]

The locations and boundaries of the sensory and motor areas of the human cortex, well known from earlier studies with different methods, have been confirmed. Imaging studies have shown, in addition, that these primary sensory areas vary in size between individuals and between hemispheres in the same brain, and that they can be changed in size by intensive sensory or motor experiences. A number of studies show that the sensory and motor areas, including primary areas, may be active when a human subject passively imagines perceptual or motor experiences, without movement.

Similarly, the location of regions active during higher-order perceptual operations (e.g., those for color and for motion processing) have been defined in the human brain. Many cognitive functions previously thought to be localized to a single cortical area have been shown to engage many, and frequently widely separated, areas of the cortex in a distributed (not hierarchical) arrangement.

Any but the simplest of cognitive operations, including language, learning, remembering, attending, and others, are also associated with more or less distinctive sets of active regions distributed over the cortex. These active nodes shift in amplitude and position in time as the execution of a cognitive task proceeds. Significantly, the areas of cortex between the active loci show no signs of task-related activation; that is to say, *there is no evidence of mass action in the cortex.* Further, the functions of the cortex of the frontal lobe in working memory, in planning, and in willing action have been confirmed, and some defects in frontal-lobe activation observed in some psychotic states.[27]

Taken together, these results sustain the related concepts of functional segregation of certain processes in local cortical regions and their functional integration into distributed system action of the several, sometimes many, local regions active during the execution of perceptual or cognitive tasks.

It should be said, however, that the results obtained until now in this rapidly developing field are geographic in nature; they provide elegant answers to geographic questions. They do not yet reveal the dynamic neuronal operations that generate the recorded changes in local blood flow. These local blood-flow changes, like the slow-wave electrical activity of the electroencephalogram, are secondary signs of neuronal activity. The next important—and difficult—step in this field is to discover how these secondary signs are related to the neuronal activities that generate them. This central problem is well known to scientists working in imaging neuroscience, and many now focus on solving it by experiments with nonhuman primates working in perceptual or cognitive tasks. The experimental paradigm includes simultaneous recording of one or more of the secondary signs of brain activity, together with samples of the relevant neuronal population activity via acutely inserted or chronically implanted microelectrodes.[28] Absent new and unpredictable discoveries or development of new methods, no other way is apparent through which the methods of observing the active human brain can be brought to a new level of productivity. A similar need is obvious for the large experimental fields of evoked potential and event-related potential recording in humans.

THE NEUROBIOLOGICAL REVOLUTION AND THE
CLINICAL NEUROLOGICAL SCIENCES

The clinical neurological disciplines, like all of medicine, have over the last half-century been transformed to an unpredicted and perhaps unimaginable degree into scientifically based enterprises. There is scarcely a field of biological or physical science that has not contributed to knowledge of brain function and how it is disordered by disease. The principle sources are molecular biology and neurobiology, cellular and systems neurophysiology, immunology, genetics, and—as evidenced by the imaging methods described above—applied physics and biomedical engineering. This is not to say that the major problems have been solved; but for many major diseases the ways in which they will be solved become clearer with each passing decade.

The most important and useful advance for the clinician has been the development and wide availability of the brain-imaging methods. They have transformed the neurological diagnosis from a Holmsian game to one of elegance and precision, especially for the location and identification of space-occupying lesions, and also for following the effects of therapy. Several research efforts in this large field of medical science merit brief description. They illustrate in contrasting ways how severely incapacitating major diseases of the brain are, and how difficult the discovery of adequate modes of treatment.

Research in biological psychiatry has been galvanized by the gains in knowledge of the molecular mechanisms of synaptic transmission in the brain. The universal working hypothesis is that these disorders are produced by abnormalities in one or more central transmitter-receptor systems; they are brain diseases. The therapeutic aim is to design drugs that will rectify that abnormality in transmitter-receptor system function by either positive or negative actions, or both. The defects are commonly attributed to genetic mutations, to the action of an undefined external agent, or some combination of the two. Research at the molocular level is complemented by studies of the behavioral abnormalities using neuropsychological methods and by definitions of the gross changes in brain structure and function frequently present. Manic-depressive illness is a debilitating familial disorder. The mechanisms of its inheritance remain uncertain, and attempts to identify the relevant genes associated with what is most likely a heterogeneous group of genetic diseases continue. Nevertheless, considerable progress has been made in achieving pharmacological control of depressive illness. The evidence is convincing that in these patients there are defects in the serotonergic (5-HT) transmitter system, and several drugs that increase 5-HT production can benefit them. Nevertheless, the suicide rate in these desperately ill patients hovers between 10 and 20 percent. Intensive searches continue for more effective pharmacological agents. Some improvements have been achieved in the therapy of other major psychiatric disorders. These advances have been obtained in a quasi-empirical fashion, in the absence of knowledge of the basic pathogenetic mechanisms whose discovery will likely lead to more effective therapies.

Parkinson's disease is a severe chronic disorder of motor control marked by tremor, rigidity, hypokinesia, and postural instability. It follows a selective degeneration of the substantia nigra, a nucleus of the brain stem, and loss of its dopaminergic innervation of the basal ganglia, particularly of that to the putamen. The disease usually begins in late middle age, and in many cases progresses over five to ten years to a severe dementia. The cause of the nigral degeneration is unknown. Treatment with the dopamine precursor L-dopa provides dramatic relief of motor defects in some patients, but its effectiveness gradually declines over five to seven years, after which there is a marked increase in the severity of the disease. This failure with time of the effectiveness of pharmacological agents has prompted determined efforts to devise other forms of treatment. Among these are attempts to replace the dopinergic action upon neurons of the putamen by transplantation into it of dopaminergic neurons obtained from other sources. Initially, homograph grafts were made from the patients' own adrenal glands (about a thousand cases worldwide) with little or no sustained beneficial effect. Subsequently, dopaminergic neurons obtained from the brain stems of human fetuses aborted at a critical stage of embryogenesis have been used. Transplantations of this sort have been made in about two hundred human patients. Some clinical success has followed this heroic effort, and several important facts have been established. A limited number (10 to 20 percent) of the transplanted cells survive, synthesize and release dopamine, and innervate the adjacent host cells of the putamen. About 10 to 20 percent of patients receiving such transplants have shown significant clinical improvement; this is rarely sufficient to allow cessation of L-dopa therapy. Further success with this method depends upon increasing the probability of transplant survival and function, and developing larger and less controversial sources of dopaminergic cells for transplantation.[29] Efforts are now being made to achieve this latter objective by means of the genetic engineering of stem cells to selectively synthesize dopamine.

Many neurological disorders are produced at least in part by inherited or acquired genetic mutations; examples include Alzheimer's disease, the major psychoses, and many other less-common disorders like the muscular dystrophies, Huntington's disease, the degeneration of long axonal tracts in the nervous system, and similar diseases. Many of these appear to be produced by multiple genetic defects; for some, genetic and other factors interact in their pathogenesis. The single relevant gene has been identified in a few neurological diseases, which brings investigators to the difficult problem of how to replace or compensate for the action of those abnormal gene products.

It has been known for some time that the brain is not "immunologically isolated" but that these two major control systems interact continuously in both health and disease. Indeed, several brain diseases are now thought to be autoimmune in nature. Of these, perhaps the most common and devastating is multiple sclerosis (MS), in which intermittently recurring inflammatory demyelinations occur widely in the central nervous system. The T-lymphocytes, an important component of the autoimmune system, are thought to play an important role in MS pathogenesis. The antigen triggering the inflammatory response has not been identified. It may be a component of "self" no longer shielded by self-tolerance, or an unknown external agent. The goals of studies of this disease are to stop the inflammatory process and to promote re-myelination. An immune therapy is sought that will achieve these ends while leaving the immune system intact. It has not yet been discovered.

On the more technical side, a host of new methods have converted neurosurgery from a heroic enterprise for both patient and surgeon to one of controlled and safe intervention. These include methods for controlling brain edema; improvements in neuroanesthesia that make intracranial procedures safe and effective; increased accuracy in diagnosis, first with cerebral angiography and then with methods of brain imaging; microsurgical techniques for successful treatment of arterial aneurysms and of occlusive vascular disease; and the combination of imaging and stereotactic methods for procedures deep within the brain.

BRAIN THEORY AND MODELING

A distinguished philosopher has written that a new theory is like a visit to a foreign country, for it reveals new things as well as old things in new relations. If so, surely the brain sciences have been well-traveled during the last two millennia, particularly in terms of theories of the mind-brain relation. Few theories proposed to solve this central problem of man's existence have yielded testable hypotheses about or contributed to knowledge of the structure and function of the brain. Nevertheless, their general trend has been of value for brain science; present-day theories have gradually shifted the focus of attention away from the idea that each man possesses an immortal soul, embodied in heart or head, to the view that the brain is a biological system, with an evolutionary history of its own. Embedded in the brain are the mechanisms that generate and control all aspects of behavior, including those at the highest cognitive levels that are commonly hidden from external observation.

The present era is not impoverished of brain theories, and the more recent theories proposed are about how the brain works. Disputations about the counterpoised themes of materialism and dualism fade into the background. Certainly one or another of the several varieties of materialism is most useful for the experimental brain scientist, who must approach myriad tasks with some theory as guide. What is desired now is to sharpen those theories in terms of what is known of neurons and the systems they inhabit. The confident expectation is that new theories will generate new and better hypotheses about brain function, and new and better experiments to test them.

Brain theory and brain models will be important features in several essays in this issue of *Dædalus*. One, based on the view of the brain as a selective system, has generated a large-scale method of neurally-based synthetic modeling that has yielded considerable success in exploring a variety of complex behaviors, particularly perceptual, at levels not yet—not *quite* yet—accessible with present experimental methods.[30]

Whether or not one favors the metaphor of the brain as a computational system matters little in view of the wealth of new and testable hypotheses generated by the union of neuroscience with computational theory, a union now little more than a decade old.[31] Computational neuroscience is an umbrella covering modeling studies that range from those simulating the functions of single nerve cells to those of global cognitive behaviors like perception, memory, learning, and attention. These models share a property differentiating them from many modeling computations in cognitive science; in each there is a primary concern with whether the structures designed and the computing operations used within them are compatible with what is known of the structure and processes of real neural networks. This imposes strong constraints on the classes of models designed and the operations within them.

Modeling of large-scale systems is now common in virtually all fields of neuroscience. Models are valuable because many important experiments in neuroscience can be designed but cannot (yet) be executed in real brains because of technical difficulties. The availability of vast computing power makes it possible to test the implications of hypotheses about any system or process under study. Good models are not descriptions, but rather include a judicious and simplifying mix of facts and informed speculation. It is important to emphasize, as many major contributors to the field of computational neuroscience have done, that the metaphor likening the brain to a standard digital computer is wholly fallacious and should be abandoned. Brains and computers differ in many ways, particularly in architecture, in the serial-processing mode in computers versus simultaneous processing in brains, and in the properties of their constituent elements: neurons can take on any one of a series of values over a continuum, transistors in digital circuits only a 0 or 1.[32]

The need for models is especially trenchant for those who wish to study actions of large numbers of neurons in the modular nodes of distributed systems of the neocortex. Simulations in models of such experiments may indicate the possible outcomes these experiments may have, though without direct proof of any, and, of equal importance, which outcomes are unlikely or forbidden. A comparison of simulation results with experimental outcomes, where experiments are possible, may lead to new experiments based upon the predictions of these models.

PERSPECTIVES

I list here some generalizing principles derived from the foregoing. They vary considerably in their ability to satisfy Berlin's criterion of being "so general, so clear," but all appear established with considerable certainty, subject always to the corrective effects of new discoveries. Almost all are products of or were fully developed by the brain research of the last half of the twentieth century. I give them in no particular order of importance or certainty, adding at once that the selection is idiosyncratic; I have surely omitted some that other scholars will deem of primary importance. My list follows:

The neuron theory.

The ionic hypothesis.

The general concept of chemically mediated synaptic transmission.

Synaptic plasticity and, from it, the plasticity of systems and maps.

The individuality of individual brains.

The modular organization of the nervous system.

The understanding of the brain's evolutionary history.

The understanding of the brain's ontogenetic history.

The view of the brain as a collection of interacting distributed systems.

The view that all mental events are brain events.

The concept that brain actions are executed by the action of large populations of central neurons; there is no evidence for a neuron theory of psychology.

The understanding that central sensory-motor linkages are in the interfaces between populations of neurons.

The view that brain excitability is controlled by widely distributed modulatory systems.

The assertion that recalled memories are constructions, not replications.

The related assertion that brain representations of external objects and events are constructions, not replications.

These principles will be expanded upon in one form or another by the articles that follow.

ACKNOWLEDGMENTS

I am grateful to Dr. Mark Bear for preparing figure 1 and the accompanying caption.

ENDNOTES

[1] Mario Bunge, *The Mind-Body Problem: A Psychobiological Approach* (New York and Oxford: Pergamon, 1980). See also Mario Bunge and Rubén Ardilla, *Philosophy of Psychology* (New York and Berlin: Springer-Verlag, 1987).

[2] See the reviews by Ilya Farber and Patricia S. Churchland, William Hirst, Marcel Kinsbourne, Edoardo Bisiach and Anna Berti, Morris Moscovitch, Robert Knight and Marcia Grabowecky, J. Allan Hobson and Robert Stickgold, and Michael Gazzaniga, all in Gazzaniga, ed., *Cognitive Neurosciences* (Cambridge Mass.: MIT Press, 1995), Section IX: Consciousness.

[3] Colin McGinn, "Can We Solve the Mind-Body Problem?" *Mind* 98 (1989): 349–366.

[4]Isaiah Berlin, *The Sense of Reality: Studies in Ideas and their History* (New York: Farrar, Straus and Giroux, 1997), 21.

[5]Alan L. Hodgkin, *The Conduction of the Nerve Impulse* (Liverpool: Liverpool University Press, 1964). See also the papers by Hodgkin and Andrew Huxley in *Journal of Physiology* 116 (1952): 449, 473, and 497; 117 (1952): 500; and 121 (1953): 403.

[6]Kenneth Stewart Cole, *Membranes, Ions, and Impulses* (Berkeley, Calif.: University of California Press, 1972).

[7]One recent study, which used imaging methods, puts the average human cranial volume at about 1,434 ml for males and about 1,325 ml for females. See P. A. Filapek et al., "The Young Adult Human Brain: An MRI-Based Morphometric Analysis," *Cerebral Cortex* 4 (1994): 344–361. For estimates of the number of neurons in the human neocortex, see J. Braegaard et al., "The Total Number of Neurons in the Human Cortex Unbiasedly Estimated Using Optical Dissectors," *Journal of Microscopy* 151 (1990): 285–304; and H. J. G. Gunderson, "Stereology of Arbitrary Particles: A Review of Unbiased Number and Size Estimates and Presentation of Some New Cases (in Memory of William R. Thompson)," *Journal of Microscopy* 143 (1986): 3–45.

[8]The neuron theory evolved from the discoveries of the Spanish neurohistologist Santiago Ramon y Cajal. For translations and interpretive comments on Cajal's work, see Javier DeFilipe and Edward G. Jones, *Cajal on the Cerebral Cortex: An Annotated Translation of the Complete Writings* (New York and Oxford: Oxford University Press, 1988).

[9]Claude Bernard discovered a century and a half ago that after muscular paralysis produced by the alkaloid curare, a muscle still contracts in response to directly applied electrical stimuli, and that conduction in the nerve axons is unimpaired. He concluded that curare acts on some special chemical entity believed to operate at the neuromuscular junction—a correct conjecture. Bernard's original work was published in *C. R. Acad. Sci. Paris* 43 (1856): 825–829.

   The history of the discovery of the chemical nature of synaptic transmission begins with the hypothesis of Elliott derived from his studies of the action of adrenaline upon denervated muscle in 1904 and continues through the discoveries of the actions of acetylcholine and adrenaline at peripheral autonomic and neuromuscular synapses by Otto Loewi, Henry Dale, and their colleagues in the period 1914–1936. See the Nobel lectures by Dale and Loewi of 1936 in *Nobelstiftelsen* [Nobel lectures], *Physiology or Medicine* (4 vols.) (Amsterdam, London, and New York: Elsevier, 1964–), vol. 2 (1922–1941). The later discovery of chemical transmission in the brain led to the development of related sciences, particularly of neuropharmacology. It is a reasonable guess that presently there are more neuroscientists working in these fields than in any other of the brain sciences, and perhaps more than all the rest together. As witness, *Medline* lists 363,890 papers published between 1966 and July of 1997 on the general subject of synaptic transmission. Of these, 121,102 are listed for the four most prevalent small molecule transmitters in the brain; for dopamine, 44,760; for acetylcholine, 31,077; for gamma-aminobutyric acid, 16,902; for norepinephrine, 28,163.

[10]These remarkable advances come from the discoveries made by Bernard Katz and his collaborators of the sequence of events in chemical transmission at the neuromuscular junction. See Katz, *The Release of Neural Transmitter Substances* (Liverpool: Liverpool University Press, 1964). Studies of synaptic transmission in the spinal cord made by Eccles and his colleagues are described in a series of papers published in the 1950s; see, for example, J. S. Coombs, John C. Eccles, and P. Fatt, "The Specific Ionic Conductances and the Ionic Movements Across the Motoneuron Membrane that Produce the Inhibitory Post-synaptic Potential," *Journal of Physiology* 130 (1955): 326–373; and Coombs, Eccles, and Fatt, "Excitatory Synaptic Actions in Motoneurons," *Journal of Physiology* 130 (1955): 374–395.

[11]These nine steps are: *docking*, in which synaptic vesicles (SVs) filled with neurotransmitter dock at the active zone opposite the synaptic cleft by an unknown targeting process; *priming*, a maturation process in which SVs become competent for fast Ca^{2+} triggered membrane fusion; *fusion-exocytosis*, in which the exocytosis of SVs is stimulated by the Ca^{2+} spike during an action potential (exocytosis requires less than 0.3 msec); *endocytosis*, in which empty SVs are rapidly internalized, and become coated; *translocation*, in which coated SVs lose their coats, and are recycled; *endosome fusion*, in which recycled SVs are fused with endosomes; *budding*, by which SVs are regenerated by budding from endosomes; *transmitter uptake*, in which SVs are re-loaded with the transmitter molecule; and finally *translocation*, by which SVs move back to the active zone, perhaps by a cytoskeleton-based transport process.

[12]Jon H. Kass and S. L. Florence, "Mechanisms of Reorganization in Sensory Systems of Primates after Peripheral Nerve Injury," *Advances in Neurology* 73 (1997): 147–158; Florence and Kass, "Large-Scale Reorganization at Multiple Levels of the Somatosensory Pathway Follows Therapeutic Amputation of the Hand in Monkeys," *Journal of Neuroscience* 15 (1995): 8083–8095; Michael M. Merzenich and W. M. Jenkins, "Reorganization of Cortical Representations of the Hand Following Alterations of Skin Inputs Induced by Nerve Injury, Skin Island Transfers, and Experience," *Journal of Hand Therapy* 6 (1993): 89–204; Michael M. Merzenich et al., "Progression of Change Following Median Nerve Section in the Cortical Representation of the Hand in Areas 3b and 1 in Owl and Squirrel Monkeys," *Journal of Neuroscience* 10 (1983): 639–665.

[13]Gerald M. Edelman, *Neural Darwinism: The Theory of Neuronal Group Selection* (New York: Basic Books, 1987); Edelman, *The Remembered Present: A Biological Theory of Consciousness* (New York: Basic Books, 1989). For a briefer and more recent version, see Edelman, "Neural Darwinism: Selection and Reentrant Signalling in Higher Brain Function," *Neuron* 10 (1993): 115–125.

[14]Daniel J. Povenelli and Todd M. Preuss, "Theory of Mind: Evolutionary History of a Cognitive Specialization," *Trends in Neuroscience* 18 (1995): 418–424; Preuss, "Argument from Animals to Humans in Cognitive Neuroscience," in Gazzaniga, ed., *The Cognitive Neurosciences*, 1227–1241.

[15]Ibid.

34   *Vernon B. Mountcastle*

¹⁶The discovery of cytoarchitecture is usually attributed to Theodor Meynert (1867), although earlier discoveries like that of the white line in the occipital cortex by Genari in 1776 hinted at what was coming. The history of the field after Meynert is studded with the names of giants in neuroanatomy: Berlin, Lewis, Campbell, Eliot Smith, Kolicker, Carl Hammarberg, Cecile and Otto Vogt, Korbinian Brodmann, and Constantin Freiherr von Economo. The list includes many distinguished living neuroanatomists.

¹⁷The original discovery of axoplasmic flow was made by Paul Weiss. See Weiss and H. P. Hiscoe, "Experiments on the Mechanism of Nerve Growth," *Journal of Experimental Zoology* 107 (1948): 315–327.

¹⁸Malcolm Young, Jack Scannell, and Gully Burns, *The Analysis of Cortical Connectivity* (Austin, Tex.: R. G. Landes & Co.; Heidelberg: Springer-Verlag, 1995); Karl Friston et al., "Functional Tomography: Multidimensional Scaling and Functional Connectivity in the Brain," *Cerebral Cortex* 6 (1996): 156–164; Daniel J. Fellerman and David C. Van Essen, "Distributed Hierarchical Processing in the Primate Cerebral Cortex," *Cerebral Cortex* 1(1991): 1–47.

¹⁹Wolf Singer and Charles M. Gray, "Visual Feature Integration and the Temporal Correlation Hypothesis," *Annual Review of Neuroscience* 18 (1995): 555–586.

²⁰Vernon Mountcastle, "The Parietal System and Some Higher Brain Functions," *Cerebral Cortex* 5 (1995): 377–390.

²¹Rodney Douglas and Kevan A. C. Martin, "A Functional Microcircuit for Cat Visual Cortex," *Journal of Physiology* 440 (1991): 735–769; Charles D. Gilbert, "Circuitry, Architecture, and Functional Dynamics of Visual Cortex," *Cerebral Cortex* 3 (1993): 373–386; Alex Thomson and Jim Deuchars, "Temporal and Spatial Properties of Local Circuits in Neocortex," *Trends in Neuroscience* 17 (1994): 119–126; Erol Basar and Theodore H. Bullock, eds., *Induced Rhythms in the Brain* (Boston: Birkhauser, 1992); J. E. Desmedt and C. Tomberg, "Transient Phaselocking of 40 Hz Electrical Oscillations in Prefrontal and Parietal Human Cortex Reflects the Process of Conscious Somatic Perception," *Neuroscience Letters* 168 (1994): 126–129.

²²There is an immense flow of monographs, textbooks, and journals in cognitive science. An excellent entry is via the three-volume set edited by Daniel N. Osherson and Howard Lasnik, *An Invitation to Cognitive Science* (Cambridge, Mass.: MIT Press, 1992). The three volumes are titled *Language, Visual Cognition,* and *Action and Thinking.* The quotation is from the introductory essay in the first volume, at xvii.

²³Berger's studies began in 1902 in attempts to record from the brains of dogs. From 1924 on he succeeded in recording the electrical activity of the human brain, using string galvanometers without amplification. From 1931–1938, with amplification, he continued the intensive studies that established him as the pioneer of the human EEG. Pierre Gloor has collected Berger's fourteen papers; see Gloor, *Electroencephalography and Clinical Neurophysiology* 18 (1969): Supplement.

[24]Paul L. Nunez, *Neocortical Dynamics and Human EEG Rhythms* (Oxford: Oxford University Press, 1995); Alan Gevins, "High-Resolution Electroencephalographic Studies of Cognition," in *Advances in Neurology* 66 (1995): 181–195.

[25]An immense literature has accumulated on brain imaging. See especially the authoritative monograph by Richard Frackowiak et al., *Human Brain Function* (New York and London: Academic Press, 1997). See also Per E. Roland, *Brain Activation* (New York: Wiley-Liss, 1993), and the 823 papers listed in Medline up to July of 1997. For more truncated descriptions, see the special issue of *Cerebral Cortex* 6 (1) (1996); D. L. Bihan and Avi Karni, "Applications of Magnetic Resonance Imaging in Human Brain Science," *Current Opinion in Neurobiology* 5 (1995): 231–237; Roberto Cabeza and Lars Nyberg, "Imaging Cognition," *Journal of Cognitive Neuroscience* 9 (1997) (1): 1–27; Verne Caviness, Jr., et al., "Advanced Application of Magnetic Resonance Imaging in Human Brain Science," *Brain and Development* 17 (1995): 399–408; M. S. Cohen and S. Y. Brookheimer, "Localization of Brain Function Using Magnetic Resonance Imaging," *Trends in Neuroscience* 17 (1994): 269–277; Karl Friston, "Imaging Cognitive Anatomy," *Trends in Cognitive Sciences* 1 (1997): 21–27; Risto Naatanen, Risto J. Ilmoniemi, and Kimmo Alho, "Magnetoenceph-alography in Studies of Human Cognitive Brain Function," *Trends in Neuroscience* 17 (1994): 389–395.

[26]For an entry into cognitive neuroscience, see the volume edited by Michael Gazzaniga, *The Cognitive Neurosciences*, and the first ten volumes of the *Journal of Cognitive Neuroscience*.

[27]David Ingvar, "'Memory of the Future': An Essay on the Temporal Organization of Conscious Awareness," *Human Neurobiology* 4 (1985): 127–136; and Ingvar, "The Will of the Brain: Cerebral Correlates of Willful Acts," *Journal of Theoretical Biology* 171 (1994): 7–12. See also Christopher D. Frith et al., *Proceedings of the Royal Society* Series B 244 (1991): 241–246.

[28]A problem for the imaging methods is the conflict between the need for spatial resolving power and time resolution. Beyond these is the basic limitation of the vascular-on time constant, and the relation of the spatial and temporal relations of the changes in blood flow to the distributions of changes in neuronal activity. The possibility exists that significant changes in neuronal population signals may occur without significant changes in overall activity, and hence no change in blood flow, e.g., changes in phase relations without changes in overall levels of activity. Amiram Grinwald and his associates are now engaged in experiments aimed at this problem. They correlate activity dependent changes in oxyhemoglobin, deoxyhemoglobin, and light scattering measured with imaging spectroscopy in the activated visual cortex with recordings of neurons in the local area with microelectrodes. Until now these have been made in anesthetized animals, in some cases through the intact dura and thinned skull. Experiments in unanesthetized monkeys are forthcoming. See Ron Frostig et al., "Cortical Functional Architecture and Local Coupling Between Neuronal Optical Imaging of Intensive Signals," *Proceedings of the National Academy of Sciences* 87 (1990): 6082–6086; Amos Arieli et al., "Coherent Spatiotemporal Patterns of Ongoing Activity Revealed by Real-Time Optical Imaging Coupled with Single-Unit Recording in Cat Visual

Cortex," *Journal of Neurophysiology* 73 (1995): 2072–2093; Dov Malonek and Amiram Grinwald, "Interactions Between Electrical Activity and Cortical Microcirculation," *Science* 272 (1996): 551–554.

[29]Ole Isacson and Terrence Deacon, "Neural Transplantation Studies Reveal the Brain's Capacity for Continuous Reconstruction," *Trends in Neuroscience* 20 (1997): 477–482.

[30]Giulio Tononi, Olaf Sporns, and Gerald M. Edelman, "A Complexity Measure for Selective Matching of Signals by the Brain," *Cerebral Cortex* 2 (1992): 310–335; ibid, *Proceedings of the National Academy of Sciences* 93 (1996): 3422–3427. See also E. D. Lumer, Gerald M. Edelman, and Giulio Tononi, "Neural Dynamics in a Model of the Thalamocortical System, I: Layers, Loops, and the Emergence of Fast Synchronous Rhythms," *Cerebral Cortex* 7 (1997): 207–227 and 228–236. Edelman has been the chief developer of the model of the brain as a selective system; see note 13.

[31]See the authoritative and entertaining monograph by Patricia S. Churchland and Terrence J. Sejnowski, *The Computational Brain* (Cambridge Mass.: MIT Press, 1992).

[32]W. W. Lytton and Terrence J. Sejnowski, "Computational Neuroscience," in Arthur K. Asbury, Gary McKhann, and W. Ian McDonald, eds., *Diseases of the Nervous System: Clinical Neurology,* 2d ed. (Philadelphia: Saunders, 1992).

Gerald M. Edelman

Building a Picture of the Brain

THE BRAIN IS AMONG the most complicated material objects in the known universe. Even before the advent of modern neuroscience, it was a commonplace that the brain is necessary for perception, feelings, and thoughts; today this is considered a truism.[1] What is the connection? Can we build a picture of the structure and dynamics of this extraordinary object that can account for the origin of the mind? This is in fact the main goal of modern neuroscience, and, when reached, it will have large consequences for humankind.

In this essay, I want to touch upon some features of the brain that make it special, features that challenge the picture of the brain as a machine. In doing so, I shall suggest alternative views to such a picture and shall range over various levels of organization of the brain, from its most microscopic structural aspects to its most abstract functions. My position shall be that the human brain is special both as an object and as a system—its connectivity, dynamics, mode of functioning, and relation to the body and the world is like nothing else science has yet encountered. This, of course, makes building a picture of the brain an extraordinary challenge. Although we are far from a complete view of that picture, a partial view is better than none at all. Before attempting a synopsis, an examination of some key features and properties of the brain is required.

Gerald M. Edelman is Director of The Neurosciences Institute and Chairman of the Department of Neurobiology at the Scripps Research Institute.

THE PRIMACY OF NEUROANATOMY

If someone held a gun to my head and threatened oblivion if I did not identify the single word most significant for understanding the brain, I would say "neuroanatomy." Indeed, perhaps the most important general observation that can be made about the brain is that its anatomy is the most important thing about it.[2]

The brain of an adult human is about three pounds in weight. It contains about thirty billion nerve cells, or neurons. The most recently evolved outer corrugated mantle of the human brain, the cerebral cortex, contains about ten billion neurons and one million billion connections, or synapses. Counting one synapse per second, we would just finish counting thirty-two million years from now. If we consider the number of ways in which circuits or loops of connections could be excited, we would be dealing with hyperastronomical numbers: 10 followed by at least a million zeros. (There are 10 followed by seventy-nine zeros, give or take a few, particles in the known universe.)

Neurons have a great variety of shapes but in general they have tree-like projections called dendrites, which receive connections. Neurons also have a single longer projection called an axon, which makes synaptic connections at the dendrites or cell bodies of other neurons. (See figure 1, plate 2.) No one has made an exact count of different neuronal types in the brain, but a crude estimate of fifty would not be excessive. The lengths and branching patterns of dendrites and axons from a given type of neuron fall within certain ranges of variation, but even in a given type, no two cells are alike.

How the brain is connected provides the major ground for understanding its general function. We know that the brain is interconnected in a fashion no man-made device yet equals. First of all, proceeding from the finest ramifications of its cells up to its major pathways, its connections are all three dimensional, or 3D. (A computer chip can be connected to other chips in 3D, but it is inscribed in 2D.) Second, a brain's connections are not *exact*. If we ask whether the connections are identical in any two brains of the same size, as they would be in computers of the same make, the answer is no. At the finest scale, no two brains are identical, not even those of identical twins. Furthermore, at any two moments, connections in the same brain are not likely to remain exactly the same. Some cells will have retracted their processes, others will have extended new ones, and certain other cells will have died. This observation applies to patterns at the finest scale, consisting of individual neurons and their synapses. Thus although the *overall* pattern of connections of a given brain area is describable in general terms, at the level of its synapses the patterns are extraordinarily complex and variable.

As an example, consider a so-called pyramidal cell in a particular layer of the six-layered cerebral cortex. It typically has as many as ten thousand synapses connecting it with distant or neighboring cells. If one moved over to the next pyramidal cell of the same type, the number of synapses could vary widely and the pattern of their contacts could be quite different. Yet in a given area of the cortex (say, that for vision), these two cells will resemble each other more than one of them would resemble a cell from another area of the cortex, for example, that for controlling movement. One conclusion we can draw from such observations is that, while there are close similarities in certain regions, there are no absolutely specific point-to-point connections in the brain. The microscopic variability of the brain at the finest ramifications of its neurons is enormous, making each brain unique. These observations provide a fundamental challenge to models of the brain based on instruction or computation. As I shall discuss, the data provide strong grounds for so-called selectional theories of the brain—theories that actually depend upon variation to explain brain function.[3]

Two key characteristics of neuronal patterns at the microscopic level are their density and their spread. The body of a single neuron measures up to about 50 microns in diameter, although its axon can range from microns to meters in length. In a tissue like the cortex, neurons are packed together at an extraordinary density; if all of them were stained with silver in the so-called Galgi stain, the section would be pitch black. (Actually, the usefulness of this stain rests on the fact that it affects only a very small fraction of cells in a given area.) Interspersed among the neurons are non-neuronal cells, called glia, that have developmental and physiological functions supporting neuroanatomy and neural activity. In some places, glia even out-number neurons.

In the dense networks of the brain, it is the spread of neuronal arbors—of dendritic trees and axonal projections—that is perhaps the most striking feature. In some places, the spatial spread of an axon forming an arbor can be over a cubic millimeter. Overlapping that arbor with all its intricate branchings can be arbors from countless other neurons. The overlap can be as great as 70 percent in three dimensional space. (No self-respecting forest, made of trees and root structures, would permit such a large overlap.) Moreover, as the axonal arbors overlap, they can form an enormous variety of synapses with cells in the paths of their branches, resulting in a pattern that is unique for each brain volume. To this day, while we can trace the full arborization of a single nerve cell, we have no clear picture of the microanatomy of the interspersed arbors of the many neighboring cells at the scale of their synapses. In summary, the main microstructural features that arrest attention are the density, the overlap, the individual branching, and the uniqueness of neuronal structures, even in the face of the quite specific higher-level patterns that characterize the neuroanatomy of a given brain region.

These larger-scale patterns and the overall functioning of the brain depend on how neurons function and exchange signals.[4] The general cellular functions of neurons—such as respiration, genetic inheritance, and protein synthesis—are like those of other cells in the body. The special features related to neural function mainly concern synapses. Neurons come in two flavors, excitatory and inhibitory, and the fine structure of their synapses varies accordingly. But for each the basic principles are similar and they involve both electrical and chemical signaling. While in certain species some synapses can be completely electrical, the vast majority of the synapses we are concerned with in human brains are chemical. In most cases, the presynaptic and postsynaptic neurons are separated by a cleft forming a single synapse. Neurons are polarized in electrical potential between the inside and outside of their cell membrane. As a result of the flow of ions across a particular portion of the cell membrane, the cell is locally depolarized. A wave of depolarization called an action potential spreads down an axon and, when it reaches the region of the synapse, causes the release of neurotransmitters from a series of vesicles in the presynaptic neuron. If the neuron is excitatory, neurotransmitters will then cross the cleft, bind to specific receptors on the postsynaptic neuron, and cause the postsynaptic neuron to depolarize. These processes occur over time periods of tens to hundreds of milliseconds. If the postsynaptic neuron depolarizes sufficiently after several such events, it will fire (i.e., generate an action potential of its own), relaying the signal in turn to other neurons to which it is connected.

A key point is that the statistics of the release of these neurotransmitters, their distribution at the microscopic level, and their binding to receptors all govern the thresholds of the response of neurons in an extraordinarily intricate and variable manner. As a result of transmitter release, electrical signaling not only takes place but also leads to changes in the biochemistry and even in the gene expression of target neurons. This molecular intricacy and the resulting dynamics superimpose several more layers of variability on that of the neuroanatomical picture, contributing to what may be called the historical uniqueness of each brain.

Despite the microanatomical variability of the brain, at its higher levels of anatomical organization it falls nicely into areas, regions, and specialized parts that, in general, are functionally segregated or specialized. In examining such regions, it is tempting to assign necessary and sufficient functions to each area. As I shall discuss below, this has historically led to paradoxes that place the local and global aspects of brain function in sharp contrast to each other. In the case of vision, for example, there are as many as three dozen (and probably more) different areas in the monkey brain, each contributing to a different function—such as the detection of line orientation, the movement of objects, or the construction of color. These areas are widely distributed over different regions of the brain, and yet there is no one master area coordinating all their functions to yield a coherent visual image or pattern. Nevertheless, coherent perceptual patterns of this kind are in fact generated by the brain.

This distributed property of different, segregated functions raises extraordinary difficulties for attempts to understand how brain anatomy relates to brain physiology. One feature found in monkey brains that affords a helpful basis for resolving these difficulties is as follows: Of the more than 305 connection paths (some with millions of axonal fibers) between members of the set of functionally segregated visual areas, over 80 percent have fibers running in both directions. In other words, the different functionally segregated areas are for the most part reciprocally connected. These reciprocal pathways are among the main means that enable the integration of distributed brain functions. They provide a major structural basis for reentrant signaling, a process that, as I shall describe later, offers the key to resolving the problem of integrating the functionally segregated properties of brain areas despite the lack of a central or superordinate area.[5]

In this abbreviated account I have not mentioned the variety of other important brain structures—such as the cerebellum, which appears to be concerned with coordination and synchrony of motion; the basal ganglia, which consist of nuclei connected to the cortex that are involved in the execution of complex motor acts; and the hippocampus, which runs along a skirt near the temporal cortex of the brain and which has a major function in consolidating short-term memory into long-term memory in the cerebral cortex. The specific ways in which these different structures interact with the cortex are of central importance, but the principles and problems connected with analyzing their functions remain similar to those I have already mentioned. (For an illustration of the general anatomy of the brain, see figure 2, plate 4.)

One of the main organizing principles of the picture we are trying to build is that each brain has uniquely marked in it the consequences of a developmental history and an experiential history. The individual variability that ensues is not just noise or error. As we shall see, it is an essential element governing the ability of the brain to match unforeseeable patterns that might arise in the future of a behaving animal. No machine we are familiar with incorporates such individual diversity as a central feature of its design. The day will certainly come, however, when we can build devices that incorporate some of the formative patterns and connection rules that we see in brains. One can safely foresee that a quite different kind of memory and learning will be exhibited by such constructions than those we currently attribute to computers.

THE SPECIALNESS OF THE BRAIN

Our quick review of neuroanatomy and neural dynamics indicates that the brain has special features of connectivity that do not seem consistent with those operating in computers. If we compare the signals a brain receives with those of such machines, we see a number of other features that are particular to brains. In the first place, the world cannot function as the unambiguous input device that a piece of computer tape is. Nonetheless, the brain can sense the environment, categorize patterns out of a multiplicity of signals, initiate movement, mediate learning and memory, and regulate a host of bodily functions.

The ability of the nervous system to carry out the perceptual categorization of various signals for sight, sound, and other sensory inputs, dividing them into coherent classes without a prearranged code, is certainly special. We do not presently understand fully how this is done, but, as I shall discuss later, I believe it arises from a special set of mapped interactions between the brain, the rest of the body, and various rearrangements of signal configurations originating in the environment.

Another distinguishing feature of the brain is how its various activities are dependent on constraints of value. I define value systems as those parts of the organism (including special portions of the nervous system) that provide a constraining basis for categorization and action within a species. I say "within a species" because it is through different value systems that evolutionary selection has provided a framework of constraints for those somatic selective events within the brain of each individual of a particular species that lead to adaptive behavior. Value systems can include many different bodily structures and functions (the so-called phenotype); perhaps the most remarkable examples in the brain are the noroadrenergic, cholinergic, serotonergic, histaminergic, and dopaminergic ascending systems. During brain action, these systems are concerned with determining the salience of signals, setting thresholds, and regulating waking and sleeping states. Inasmuch as synaptic selection itself can provide no specific goal or purpose for an individual, the absence of inherited value systems would simply result in dithering, incoherent, or nonadaptive responses. Value constraints on neural dynamics are required for meaningful behavior and for learning.

If we consider neural dynamics, the most striking special feature of the higher vertebrate brain is the occurrence of the process I have called "reentry."[6] Reentry depends on the possibility of cycles in the massively parallel graphs of the brain, such as are seen in their most elementary form in reciprocally connected brain maps. It is the ongoing recursive dynamic interchange of signals occurring in parallel between maps that continually interrelates these maps to each other in space and time. If I were asked to go beyond what is merely special and name the *unique* feature of higher brains, I would say it is reentry. There is no other object in the known universe so completely distinguished by reentrant circuitry as the human brain. While a brain has similarities to a large ecological entity like a jungle, no jungle shows anything remotely like reentry, nor do human communication systems; reentrant systems in the brain are massively parallel to a degree unheard of in our communication nets, which, in any event, deal with coded signals.

All of these special features of the brain—the connectivity, the categorizing ability, the dependence on value, the reentrant dynamics—operate in a very heterogeneous fashion to yield coordinated behavior. As I hinted above, the nonlinear aspects of the interaction between the brain, the body, and various parallel signals from the environment must be considered together if we are to understand categorization, movement, and memory.

Finally, there is also something quite special about the brain when considered from an evolutionary point of view. While the human brain can be viewed in terms of the immense anatomic, biochemical, and dynamic complexity that is built up during individual development and behavior (a complexity that includes an enormous variety of different events), its actual evolution can only be explained by a relatively small number of selectional events or transcendences.[7]

Understanding the specialness I have touched upon here sets a good part of the research agenda for modern neuroscientists, who are attempting to build a picture of the brain and thus understand its functions. This understanding and its research agenda require a robust theory—one that provides insights into the biological origins of pattern formation, the nature of complexity, and the correlation of brain activity with psychological functions. I turn now to a brief description of one such theory.

NEURAL DARWINISM

The diversity and individuality of the multilayered structures and dynamics of each brain pose major challenges to the formation of any theory proposed to account for global brain function. Machine analogies simply will not do. I believe that these challenges can be met by turning to population thinking. Charles Darwin invented population thinking, the idea that variation in individuals of a species provides the basis for the natural selection that eventually leads to the origin of other species. Darwin's description of the combined processes of variation and selection in populations has since been shown to be the most fundamental of biological principles. This population principle not only provides the basis for the origin of species but also governs processes of somatic selection occurring in individual lifetimes. An example is the immune system, in which the basis for molecular recognition of foreign molecules is somatic variation in antibody genes, leading to a vast repertoire of antibodies with different binding sites. Exposure to a foreign molecule is followed by the selection and growth of cells bearing just those antibodies that fit a given foreign structure sufficiently well, even one that has never occurred before in the history of the earth. The mechanisms and the timing differ in evolution and immunity but the principles are the same—the Darwinian processes of variation and selection.

Almost two decades ago, I began to think about how the mind could arise in evolution and development.[8] It seemed to me then, as it does now, that the mind must have arisen as a result of two processes of selection: natural selection and somatic selection. The first is hardly doubted except perhaps by some philosophers and theologians. Thinking about the second led to the proposal of a theory concerned with the evolution, development, structure, and function of the brain.

This theory of neuronal group selection (TNGS), or neural Darwinism,[9] has three main tenets: 1) *Developmental selection.* During the development of a species, the formation of the initial anatomy of the brain is constrained by genes and inheritance but connectivity at the level of synapses is established by somatic selection during an individual's ongoing development. This generates extensive variability of neural circuitry in that individual and creates groups made up of different types of neurons. Neurons in a group are more closely connected to each other than to neurons in other groups. 2) *Experiential selection.* Overlapping this early period and extending throughout life, a process of synaptic selection occurs within the diverse repertoires of neuronal groups. In this process, certain synapses within and between groups of neurons (locally coupled neurons) are strengthened and others are weakened. This selectional process is constrained by value signals that arise as a result of the activity of ascending systems in the brain, an activity that is continually modified by successful output. 3) *Reentry.* Spatiotemporal correlation of selective events in the various maps of the brain is mediated through the dynamic process of reentry. This occurs via the ongoing activity of massively parallel reciprocal connections, and it correlates events of synaptic selection across disjunct brain maps, binding them into circuits capable of coherent output. Reentry is in fact a form of ongoing higher-order selection, a form that appears to be unique to animal brains.

Because of the dynamic and parallel nature of reentry and because it is a process of higher-order selection, it is not easy to provide a metaphor that captures all of its properties. But as an example, imagine a string quartet in which each player responds through improvisation to ideas and cues of his own as well as to sensory cues of all kinds in the environment. Since there is no score, each player decides what notes to play and so is not coordinated with the other players. Now imagine that the players are connected to each other by myriad threads so that their actions and movements are very rapidly conveyed by back-and-forth signals. Signals that instantaneously connect all four players would lead to the correlation of their sounds, and thus a new, more cohesive, and integrated sound would emerge out of the independent efforts of each player. This correlative process would alter the next action of each player and the process would be repeated but with new emergent and more correlated melody lines. Although no conductor would instruct or coordinate the group, the players' productions would tend to be more integrated and more coordinated, leading to a kind of music that each one alone could not produce.

One of the striking consequences of the process of reentry is the emergence, as with the players in our example, of widespread interactions among different groups of active neurons distributed across many different and functionally specialized brain areas. The resultant spatiotemporal correlation of the activity of widely dispersed neurons is the basis for the integration of perceptual and motor processes, which is characterized by the global coherency and unified character of such processes. Indeed, if reentrant paths connecting cortical areas are disconnected, these integrative processes are disrupted. Reentry allows for a unity of perception and behavior that would otherwise be impossible given the absence in the brain of a unique central processor with detailed instructions or algorithmic calculations for coordinating functionally segregated areas.

Following a brief account of this theory in 1976, I published a trilogy between 1987 and 1989 describing its various aspects ranging from neuroanatomical development to consciousness.[10] In addition, along with my colleagues, I published a series of papers that analyzed various aspects of the theory, describing a number of simulations and models that tested its self-consistency.[11] Since that time considerable evidence to support the theory has accumulated, and no disconfirming evidence has emerged.[12] Nevertheless, it has become obvious that much remains to be clarified and explained in greater depth. One of the most important issues concerns analyzing and understanding the complexity of the brain that underlies its integrative capacities.

COMPLEXITY AND ITS DISCONTENTS

The ideal of science is to provide the simplest account possible of the facts of the world. As Einstein put it, a theory must be as simple as possible but no simpler. The fact is that the world of biology and particularly that of the brain, not to speak of the world of human experiences, leaves us almost stupefied with complex chains of events and at the same time hardly satisfied with naive attempts at simple reduction. I began this essay with the statement that the brain is among the most complicated natural objects in the known universe. This is both a challenge and a source of discontent; it does not appear that the brain will yield itself to simple mathematical reduction. What one would like to have, nevertheless, is a measurable characterization of the brain's complexity that clarifies the relations between its functions and its structure. This requires the creation of a yardstick or measure of complexity that captures the essential elements of the brain's connectivity and dynamic interactions.

What is complexity? Ideas of heterogeneous mixtures, of multiple levels and scales, of intricate connectivities, and of nonlinear dynamics spring to mind.[13] One approach to relate these matters is in terms of the degree of independence and mutual dependence of the functions of the brain's multiple elements. This provides a means to define a complexity measure for the brain: It turns out that, using such an approach, we can apply the concepts of statistical entropy and mutual information that have been successfully applied to the analysis of multivariate processes. I will attempt here only to give a qualitative feeling for the quantitative approach that my colleagues have applied to derive various measures of complexity.[14]

If the components of a system are independent, entropy (which is a measure of disorder) is maximal and the mutual information between the system's parts is zero. If the system has any constraints resulting from interactions among its parts, however, they will deviate from statistical independence; mutual information will in general increase, and entropy is reduced. To provide an intuitive feel for a complexity measure based on entropy and mutual information, consider that it would have a low value for an ideal gas, which has statistically independent components (high entropy), as well as for a perfect crystal that has totally integrated units (low entropy but no local segregation or functional segregation) and in which all regions resemble all others.[15] This is certainly not generally true of brains—brains are functionally segregated into a diversity of different units and regions essential for different tasks. The derivation of complexity measures rests on reconciling this diversity with the global exchange and overall integration of different segregated areas.[16]

A word about functional segregation in the brain is in order. During development and behavior, neuronal groups are formed that consist of local collectives of strongly interconnected neurons that share inputs, outputs, and response properties. Each group is connected to only certain subsets of other groups and possibly to particular sensory afferents or motor efferents. In a given brain area, different combinations of groups are activated preferentially by different input signals. In the visual cortex, for example, at the level of brain maps there is a functional segregation into areas, each responding to different stimulus attributes such as color, motion, or form. Functional segregation is clearly revealed by the appearance of specific perceptual and motor deficits that result from lesions to particular cortical areas.

A sharp contrast to this kind of specialization is provided by equally strong evidence for global integration. This occurs across many levels, ranging from interactions among neurons to interactions among areas to concerted outputs that lead to particular behaviors. Integration results jointly from the patterns of connectivity among neurons and their dynamic interactions. Any two neurons in the brain are separated from each other by a relatively small number of synaptic steps. Moreover, the pathways linking the functionally segregated neuronal groups are often reciprocal. This reciprocity provides the necessary substrate for reentry, which I have already described as a central integrative process of ongoing recursive signaling occurring along multiple, massively parallel paths. Simultaneous activity in subsets of these paths allows for selection and correlated activity among neuronal groups in the same and different areas and even between distant mapped areas.

Although much evidence exists for both localization of function and for global processes, conflicts in interpreting this evidence have given rise to long-standing controversies in brain science. To explain brain function, "localizationists" favor specificity of local brain modules whereas "holists" stress global integration, mass action, and Gestalt phenomena. As is often the case, when viewed from the proper vantage point, these controversies turn out to be misguided. This vantage point is provided by formulating a framework for an analysis of brain complexity that suggests that effective brain function arises both from the combined action of local segregated parts having different functions and from the global integration of these parts mediated by the process of reentry.

A brain organized in this fashion is not just a complex system, it is a complex selectional system. It consists of a series of repertoires—variants of neural circuits that are selected by interactions with inputs and that engage in signaling to outputs. Various signals from the environment provide sensory input to such repertoires, and the selection of appropriate neural circuits results in a matching or a fit to that input. The complexity of repertoires is in fact an essential property in the success of this process. After repeated episodes of selection, a complex neural system can generally come to match the overall statistical structure of signals from the environment by changing which paths are favored.

At any one time, an individual input or stimulus inevitably contains only a very small subset of the regularities that are potential in this statistical structure. Nonetheless, after selection, a given stimulus that is consistent with the overall statistical structure of previous inputs will tend dynamically to enhance the set of intrinsic correlations already present in the neural system. Thus, the functional connectivity of the brain and the presence of memory, which I shall discuss below, provide an intrinsic context that by necessity dominates the brain's dynamic response to any single stimulus. This is perhaps not surprising given the fact that, in mammalian brains, most neurons receive signals from other neurons, rather than directly from sensory inputs. The reentrant recursive signaling among multiple sets of neuronal groups across massively parallel reciprocal connectivity assures that such an intrinsic context is made available in a rapid fashion in very short time periods. A set of examples of this process may be seen in a series of functioning computer models of the cerebral cortex.[17]

By such means, a complex brain that has undergone neuronal group selection and reentrant interactions can go beyond the information given—it can generalize, fill in ambiguous signals, and generate various sensory constancies. The synaptic changes that are essential to these processes are connected to the emergence of memory, one of the most fundamental properties of neural systems.

NONREPRESENTATIONAL MEMORY: A SYSTEM PROPERTY

To say, as is commonplace, that memory involves storage raises the question: What is stored? Is it a coded message? When it is "read out" or recovered, is it unchanged? These questions point to the widespread assumption that what is stored is some kind of representation. This in turn implies that the brain is supposed to be concerned with representations, at least in its cognitive functions. In perception, for example, even before memory occurs, alterations in the brain are supposed to stand for, symbolize, or portray what is experienced. In this view, memory is the more or less permanent laying down of changes that, when appropriately addressed, can recapture a representation—and, if necessary, act on it. In this view, learned acts are themselves the consequences of representations that store definite procedures or codes.

The idea that representational memory occurs in the brain carries with it a huge burden. While it allows an easy analogy to human informational transactions embedded in computers, that analogy poses more problems than it solves. In the case of humans working with computers, semantic operations occurring in the brain, not in the computer, are necessary to make sense of the coded syntactical strings that are stored physically in the computer either in a particular location or in a distributed form. Coherency must be maintained in the code (or error correction is required) and the capacity of the system is quite naturally expressed in terms of storage limits. Above all, the input to a computer must itself be coded in an unambiguous fashion; it must be syntactically ordered information.

The problem for the brain is that signals from the world do not in general represent a coded input. Instead, they are potentially ambiguous, are context-dependent, are subject to construction, and are not necessarily adorned by prior judgments as to their significance. An animal must categorize these signals for adaptive purposes, whether in perception or in memory, and somehow it must associate this categorization with subsequent experiences of the same kinds of signals. To do this with a coded or replicative storage system would require endless error correction, and a precision at least comparable to and possibly greater than that of computers. There is no evidence, however, that the structure of the brain could support such capabilities directly; neurons do not do floating-point arithmetic. It seems more likely that such mathematical capabilities have arisen in human culture as a consequence of symbolic exchange, linguistic interactions, and the application of logic.

Representation implies symbolic activity. This activity is at the center of our semantic and syntactical skills. It is no wonder that, in thinking about how the brain can repeat a performance, we are tempted to say that the brain represents. The flaws with such an assertion, however, are obvious: there is no precoded message in the signal, no structures capable of high-precision storage of a code, no judge in nature to provide decisions on alternative patterns, and no homunculus in the head to read a message. For these reasons, memory in the brain cannot be representational in the same way as it is in our devices.

What is it then, and how can one conceive of a nonrepresentational memory?[18] In a complex brain, memory results from the selective matching that occurs between ongoing neural activity and signals from the world, the body, and the brain itself. The synaptic alterations that ensue affect the future responses of the brain to similar or different signals. These changes are reflected in the ability to repeat a mental or physical act in time and in a changing context. It is important here to indicate that by the word "act," I mean any ordered sequence of brain activities in a domain of perception, action, consciousness, speech, or even in the domain of meaning that in time leads to neural output. I stress time in my definition because it is the ability to recreate an act separated by a certain duration from the original signal set that is characteristic of memory. And in mentioning a changing context, I pay heed to a key property of memory in the brain: that it is, in some sense, a form of *recategorization* during ongoing experience rather than a precise replication of a previous sequence of events.[19]

What characteristics must the brain possess to show memory without coded representation? Just those characteristics, I believe, that one would find in a selectional system. These are a set of mappings making up a diverse repertoire, a means of changing the population characteristics of the repertoire with varying input signals, and a set of value constraints acting to enhance adaptation by repeating an output. Signals from the world or from other brain parts will act to select particular circuits from the enormously various combinatorial possibilities available in a given brain area. Selection occurs at the level of synapses through alteration of their efficacy. Which particular synapses are altered depends on previous experience as well as upon the variously combined activities of ascending value systems such as the locus coeruleus, raphé nucleus, and cholinergic nuclei.

The triggering of *any* such set of circuits resulting in a set of output signals that are sufficiently similar to those that were previously adaptive provides the basis for a repeated mental act or physical performance. In this view, a memory is dynamically generated from the activity of certain selected subsets of circuits. These subsets are degenerate in the sense that comparison would indicate that different ones contain circuit topologies or patterns that are nonisomorphic; nevertheless, any one of them can yield a repetition of some particular output. Under these conditions, a given memory cannot be identified uniquely with any single specific set of synaptic changes. This is so because the particular set of changed synapses that leads to a given output, and eventually to a performance at a given time, is itself continually changing. So it must be the adequate *pattern* underlying a performance, not its detail, that is repeated.

We conclude that synaptic change is essential to memory but is not identical to it. There is no code, only a changing set of circuit correspondences to a given output. The equivalent members of that set can have widely varying structures. It is this property of degeneracy that allows for changes in particular memories as new experiences and changes in context occur. Memory in a degenerate selectional system is recategorical, not strictly replicative. There is no prior set of determinant codes governing the categories, only the previous population structure of the network, the state of the value systems (which interact combinatorially according to context), and the physical acts carried out at a given moment. The dynamic changes linking one set of circuits to another within the enormous graph set of neuroanatomy allows the brain to create a memory. By these means, structurally different circuits within the degenerate repertoires are each able to produce a particular output leading to repetition or variation of a given mental or physical act. This property underlies the well-known associative properties of memory systems in the brain, for each member of the degenerate set of circuits has different network connections.

In this view, there are many hundreds, if not thousands, of separate memory systems in the brain. They range from all of the perceptual systems in different modalities to those systems governing intended or actual movement to those of the language system and speech sound. This gives recognition to the various types of memory tested by experimentalists in the field—procedural, semantic, episodic, and so on—but it does not restrict itself only to these types, which are defined mainly operationally and to some degree biochemically.

While individual memory systems differ, the key general conclusion is that memory itself is a system property. It cannot be equated solely to circuitry, synaptic changes, biochemistry, value constraints, or behavioral dynamics. Instead, it is the dynamic result of the interactions of *all* of these factors within a given system acting to select an output that repeats a performance. The overall characteristics of a particular performance may be similar to a previous performance within some threshold criterion, but the structures underlying any two similar performances can be quite different.

Besides guaranteeing association, the property of degeneracy also gives rise to the robustness or stability of memorial performance. There are large numbers of ways of assuring a given output. As long as a sufficient population of subsets remains to give an output, neither cell death, nor intervening variables competitively removing a particular circuit or two, nor switches in contextual aspects of input signals will, in general, be sufficient to extirpate a memory.

It might be argued that, even in a selectional system, the entire set of all the responses that give a repeated performance can be considered to be a representation. To accept this, however, would tend to weaken the notion of selection, which is a dynamic one. Selection is ex post facto; no code or symbol stands for a given memory, and very different structures and dynamics can give rise to the same memory. Above all, the structures underlying memory change continually over time. It seems senseless to conflate such dynamic properties with those that we know are characteristic of the coded representational systems, which we have consciously constructed under a code for human communication and cultural purposes.

It may be illuminating to try to envision the operation of a nonrepresentational memory by using an analogy. The geological example I use here is somewhat trivial and in only four dimensions, but I believe it can be generalized to higher dimensions. Consider a mountain, with a small glacier at its top, under changing climatic conditions or contexts leading to melting and refreezing. Under one set of warming conditions, a certain set of rivulets will merge downhill to a stream that feeds a pond in the valley below. Let that be the output leading to a repeated performance, i.e., the thawing of ice and its flow into the pond has occurred before. Now change the sequence of weather conditions, freezing some rivulets then warming them, leading to a merger with some other rivulets and the creation of new ones. Even though the structure at the heights is now changed, the same output stream may be fed exactly as before. But given even a small further change in the temperature, wind, or rain, a new stream may result, feeding or creating another pond, which may be considered to be associated with the first. With further changes, the two systems may merge rivulets and simultaneously feed both ponds. These may in turn become connected in the valley.

Consider the value constraints figuratively to be gravity and the overall valley terrain, the input signals to be the changes induced by contextual alterations of weather, the synaptic change to involve freezing and melting, the detailed rocky pattern down the hill to be the neuroanatomy, and you have a way of repeating a performance dynamically without a code. Now switch and imagine the vast set of graphs constituted by the actual neuroanatomy of the brain and consider that their connections define an n-dimensional space. By extension of the dimensionality of the process I have described, one can at least figuratively see how a dynamic nonrepresentational memory might work.

Such a memory has properties that allow perception to alter recall, and recall to alter perception. It has no fixed capacity limit since it actually generates "information" by construction. It is possible to envision how it could generate semantic capabilities prior to syntactic ones. It is robust, dynamic, associative, and adaptive. If such a view is correct, every act of perception is to some degree an act of creation and every act of memory is to some degree an act of imagination. Biological memory is creative and not strictly replicative.

A final word on the general significance of biological memory. Elsewhere, I have suggested that the ability to repeat a performance with variation under changing contexts first appeared with self-replicating systems under natural selection.[20] The action of natural selection then gave rise to various systems for which memory is critical, each having different structures within a given animal species—structures that range from the immune system to reflexes and even to consciousness. In this view, there are as many memory systems as there are systems capable of autocorrelation over time, whether constituted by DNA itself or by the phenotype that it constrains. Morphology underlies the particular properties of a given memory system. In turn, memory is a system property, allowing the binding in time of selected characteristics having adaptive value. If symmetry is a great binding principle of the physical universe, memory in selectional systems may be seen as a great binding principle in the biological domain.

CONSCIOUSNESS: THE REMEMBERED PRESENT

The greatest challenge to modern neuroscience is to provide an adequate explanatory basis for consciousness.[21] My purpose here is not to review the various attempts to provide such a basis. Instead, I want to consider a proposal for the neural origin of consciousness consistent with selectional brain theories. It is important to point out some limits on such an enterprise. The first has to do with evolution. Consciousness depends on a certain morphology. Insofar as that morphology is the product of evolutionary selection, consciousness is also such a product. At the outset of any sensible scientific consideration of consciousness, we must confront the peculiarities and limitations that arise from these conclusions.

The only morphology providing a secure functional basis for scientific assertions about consciousness is that of our own species. There are two reasons for this. First, each of us knows what it is like to be conscious as human beings but not directly what it is like to be conscious as some other animal might be; and second, we can use linguistic exchanges together with objective scientific examinations to confirm causes or correlations related to consciousness in a way that other animals cannot. The first reason recognizes that, as a process, consciousness occurs in individuals (and in the human case, in selves or persons). This reflects the fact that, in its fullest expression, consciousness is epistemically subjective: all exhaustive and historical accounts of an individual's consciousness can be provided only by that individual and cannot be directly shared or experienced by any other. The second reason recognizes that even the most parsimonious scientific (i.e., epistemically objective) account of consciousness can only be useful if our measurements and models are accompanied by correlation with reportable subjective states.

Having said that our best bet is in investigating consciousness first in human beings, we must still consider the criteria by which other animals may be investigated. Here, though, there will inevitably be additional methodological limits, and certain familiar cautions against conflating analogy and homology are bound to arise. At the anatomical level, we are on reasonably sound ground—it is known, although perhaps not exhaustively, which structures in the human brain are necessary and sufficient for consciousness. The presence of such structures in a nonhuman animal that demonstrates behavior indicative of the exchange of signs or of symbolic reference provides at least some justification for the working hypothesis that that animal is conscious. The *absence* of such structures will not allow us to say dogmatically that a particular organism is *not* conscious, but on evolutionary and functional grounds we may surmise that whatever phenomenal experience that organism has, it will not resemble ours. Given the other methodological limitations, the matter must be left at that.

Consciousness is a subjective state of sentience, unique in each subject but with properties shared by different subjects. It is implicitly felt by us as much by the recall of its absence (in deep sleep, in drugged or anesthetic states, or post-traumatically) as it is by its presence in various degrees of awareness. William James stressed that it is a process that is personal, continuous but ever-changing, selective in time, dealing mainly but not exclusively with objects independent of the self, and not exhaustive of the objects with which it deals. A predominant property of consciousness, involving referral to objects, has been termed "intentionality" by Brentano. But note that intentionality is not always present—one can be aware of a mood without any reference to objects.

There is in most states of consciousness some kind of location in time and space that is not necessarily coherent, as dreams attest. There is usually the experience of mood, and above all of so-called qualia—the subjective experiencing of the sensory modalities of sight, hearing, touch, olfaction, and taste, proprio-ception, and kinesthesia. And, of course, there are what might be called the "super qualia"—the philosopher's exemplary companions, pleasure and pain.

A biological theory of consciousness must describe how such properties arise in terms both of ongoing neural structure and function and of evolutionary events. In formulating such a description, it is useful to distinguish primary consciousness and higher-order consciousness. Primary consciousness is seen in animals with brain structures similar to ours (such as dogs) that appear able to construct a mental scene but seem to have very limited semantic or symbolic capabilities and no true language. Higher-order consciousness (which presumes the coexistence of primary consciousness) is accompanied by a sense of self and the ability to construct past and future scenes in the waking state. Higher-order consciousness requires, at the minimum, a semantic capability and, in its most developed form, a linguistic capability.

What structures and mechanisms must be described to account for the consciousness that we ascribe to dogs and to ourselves when, in certain of our subjective states, we are least in bondage to language? Here we must face four complex processes and their interactions. The first is a property shared by all animals—perceptual categorization, the ability to carve up the world of signals into categories useful for a given phenotype in an environment that has physical constraints but which itself contains no such categories. Along with the control of movement, perceptual categorization is the most fundamental process of the vertebrate nervous system. I have suggested that it occurs in higher vertebrates as a result of reentrant signaling among mapped and unmapped brain areas.[22] It occurs, usually simultaneously, in a number of modalities (sight, hearing, proprioception, et cetera) and in a variety of submodalities (color, orientation, and forms of motion, for example).

The second process required for the understanding of primary consciousness concerns the development of concepts that provide the ability to combine different perceptual categorizations related to a scene or an object, and thus to construct a "universal" reflecting the abstraction of some common feature across a variety of percepts. I have proposed that concepts arise from the mapping of the activities of the brain's various areas and regions by the brain itself.[23]

The third and fourth processes contributing to the emergence of primary consciousness are those related to memory and value. According to the theory of neuronal group selection, memory is the capacity specifically to recategorize, or to repeat or suppress, a mental or physical act. That capacity arises from combinations of synaptic alterations in reentrant circuits. As a result of categorical selection biased by value systems to yield such synaptic change, the entire sensorimotor system is *constrained* to give a particular range of outputs for a particular combination of inputs.

Inasmuch as a selectional nervous system is not preprogrammed, it requires such value constraints to develop categorical responses that are adaptive within a species. Certain value systems are specifically adapted to signal salience, as can be seen in the workings of the diffuse ascending systems of the brain. Such systems—the locus coeruleus, raphé nucleus, cholinergic nuclei, histaminergic systems in the posterior hypothalamus, and various dopaminergic systems—may interact combinatorially to signal salience after receiving some particular signal sequence or after instituting an action.

Value systems of these kinds are richly connected to the concept-forming regions of the brain, notably the frontal and temporal cortex but also the parietal regions and hippocampus. These regions affect the dynamics of memories that, in turn, are established or not, depending upon positive or negative value responses. The synaptic alterations that combine to develop such a "value-category memory" are essential to a model of primary consciousness.

With the notions of perceptual categorization, concept formation, and value-category memory in hand, we can formulate a model of primary consciousness.[24] The model assumes that, during evolution, the cortical systems leading to perceptual categorization were already in place before the connections that led to primary consciousness appeared. With further development of secondary cortical areas, conceptual memory systems appeared. At a point in evolutionary time corresponding roughly to the transition between reptiles branching to birds and to mammals, a new anatomical connectivity appeared. Massively reentrant connectivity arose between the multimodal cortical areas carrying out perceptual categorization and areas responsible for value-category memory. The key candidate structures for this connectivity responsible for the emergence of consciousness are the specific thalamic nuclei, modulated by the reticular nucleus in their reentrant connectivity to cortex, the intralaminar nuclei of the thalamus, and the grand systems of corticocortical fibers.

The dynamic reentrant interactions mediated by these connections occur within time periods ranging from hundreds of milliseconds to seconds—the "specious present" of William James. What emerges from these interactions is an ability to construct a scene. Through corticothalamic reentry, the ongoing parallel input of signals from different sensory modalities in a moving animal results in a grouping of perceptual categories related to objects and events. Their salience is governed in that particular animal by the activity of value systems. This activity is influenced by that animal's history of rewards and punishment accumulated during its past behavior. The ability of an animal to connect events and signals in the world (whether they are causally related or merely contemporaneous), and thereby to construct a scene that is related *to its own* value-category memory system, is essential for the emergence of primary consciousness.

The ability to construct such a scene is the ability to construct, within a time window of fractions of seconds up to several seconds, a remembered present. An animal without such a system could still behave and respond to particular stimuli and, within limits, even survive. But it could not link events or signals into a complex scene, constructing relationships based on its own unique history of value-dependent responses. It could not imagine scenes and evade complex dangers. It is this ability to imagine that underlies the evolutionarily selective advantage of primary consciousness.

How can such a picture be reconciled with our empirical program of starting with the human experience of consciousness? To confront this issue, we must consider the later evolutionary appearance of brain structures leading to higher-order consciousness. An animal having primary consciousness alone can generate a "mental image" or a scene. This is based in part on immediate multimodal perceptual categorization in real time, and is determined by the succession of real events in the environment. Such an animal has biological individuality but no concept of self and, while it has a "remembered present," it has no concept of the past or future. These are characteristics that emerged in evolution when semantic capabilities appeared, perhaps earliest in the precursors of hominids. When linguistic capability appeared in the precursors of *Homo sapiens,* higher-order consciousness flowered. Syntactical and semantic systems provided a new means for symbolic construction and memory. As in the case of primary consciousness, a key step in evolution was again the development of the neural substrate that permits the brain to construct another novel reentrant connectivity, this time between the symbol-based sensorimotor memory systems for language and the rest of the brain.

The emergence of speech and of these new connections allowed reference to objects or events between two or more early humans or protohumans under the auspices of a symbol. The development of a lexicon of such symbols, probably initially based on the nurturing and emotive relationships between mother and child, led to the discrimination of individual consciousness and the emergence of a self. And with narrative capabilities linked to linguistic and conceptual memory, the emergent higher-order consciousness could foster the development of concepts of the past and future related to that self and to others. At that point, the individual self is freed to some extent from bondage to the remembered present. If primary consciousness marries the individual to real time, higher-order consciousness allows for a kind of occasioned divorce made possible through the creation of concepts of time—concepts of the past and the future. A whole new world of intentionality and awareness, of categorization and discrimination can be experienced and remembered by these means. Extraordinary powers are rendered by these evolutionary developments; as a result, concepts and thinking can both flourish.

The building up of a system of higher-order consciousness capable of rehearsal and planning based on value-laden memory and linguistic capabilities is accompanied by phenomenal experience and feeling. The neural processes underlying that feeling can have far-reaching causal consequences—as we know by witnessing the concomitants of pain or pleasure. Feelings or sensations are the identifiers to a conscious individual of particular sets of neural states. The discussion so far has not addressed the origins and basis for feelings themselves. This issue of qualia is considered by some to be the sticking point and major obstacle to any theory that attempts to explain consciousness as a causal consequence of neural events. I have pointed out, however, that a scientific theory of consciousness, like any other scientific theory, implies intersubjective communication between at least two human beings. For that communication to be scientifically successful, we must already assume that the prior transactions of each of the two communicants must allow both to experience qualities: warm, green, rough, and so on. These are aspects of qualia, and thus it may seem we are in a circular situation: to explain qualia scientifically, we must assume their existence. Our theory proposes, however, that embodiment is a source of meaning and that experiences leading to meaning certainly involve qualia—to be conscious is to experience qualia. There is no qualia-free human observer; actual scientific observers must have sensations as well as perceptions. There is thus no God's-eye view of consciousness that would succeed in conveying or explaining what "warm" is to a hypothetical, qualia-free observer. Description, scientific or otherwise, must not be confused with embodiment: being is not describing. Despite this limitation, we can account scientifically for the discriminability of qualia and for their refinement as species of higher-order categorizations that are carried out by complex reentrantly connected brains.[25]

If indeed there is one central organizing principle underlying the appearance of consciousness it is the emergence of a few specific but very critical reentrant systems in evolution. These served to relate new forms of memory to perceptual and conceptual activities in the brain under the constraint of values. Following the emergence of primary consciousness, the development of linguistic capabilities through new reentrant connections of the language areas of the brain to various distributed areas mediating concept formation led to higher-order consciousness, which flowered in humans. Unraveling the complex of neural connections (mainly in the thalamocortical system) that gave rise to these extraordinary processes remains as a central challenge for modern neuroscience.

THE PICTURE OF THE BRAIN: TWO EXCLUSIVE VIEWS

In considering how we may build a picture of the brain I have deliberately emphasized one view, in full recognition of the fact that it is not the most widespread or received view. Indeed, at least by contrast, it might be called a radical view.[26]

It may be useful to highlight the contrast between the received and the radical views to indicate how differently our factual picture of the brain is presently interpreted. In the received view, the brain is an information machine with precise circuit functions and mechanisms to react to and store signals in a more or less orderly and coded fashion. In this view, anatomy involves precise connectivity with high specificity. If variation exists, it is error, or else it is ignored in the focus on precise circuit functions. Physiology, according to the received view, consists of highly defined circuit functions governed by a form of intensity or amplitude coding in which synaptic change mediates the storage of information in order to call up coded specific circuits later for specific functions. The brain is considered to carry out computations and, indeed, is felt by many to be a special kind of computer. It is implicit in this view that input to the brain from the world is somehow syntactically parsed, as in a computer.[27]

In the received view, the world presents *information* to the brain in the sense that the ordering of input signals follows certain objective categories of grouping that go beyond mere energy differences between neighboring sources of signals. Memory involves the storage of this information, presumably via a variety of codes at synapses. Recall involves calling up patterns of such stored codes. The purpose of memory and learning is to yield coordinated output according to appropriately coded routines. The various functions of perception, sensation, movement, attention, sleep, and consciousness are integrated in much the same manner as they would be if they were programmed in a computer. While, like all summaries, this one is to a degree overdrawn, I believe its flavor is right.

What about the radical view, one version of which I have emphasized in this essay? According to this view, the brain is a dynamic system that emerges from a selectional interaction with the world. The selection is constrained by evolved value, not by instructions or codes, and is actually determined by sensorimotor encounters with a world of signals. This world does not contain unambiguously parsed cues but acts with the brain to permit construction of adaptively valuable behavior. If one could look inside an appropriately selected set of brain repertoires and could read the translation to and from the phenotype, one would see an enhanced capacity for certain complex dynamics but no code. Thus one would not see uniquely mapped representations.

In the radical view, anatomy is a composite made up of patterns with underlying variability at the finest ramifications of neural connectivity. This variability and local diversity is looked upon not as noise, potentially leading to error in a code, but rather as an essential substrate for selection—what is called the primary repertoire, made up of highly diverse circuits in a given brain area.[28] A key element of neuroanatomy is the existence of massive reciprocal connectivity, which provides the structural basis for reentry. A major emphasis is placed on degeneracy, the existence of many nonisomorphic pathways that can nonetheless lead to similar functional outcomes.

At the level of gross physiology, circuits are functionally segregated parts that have significance only in the context of the action of very large correlative networks. Synaptic change within these networks is a reflection of selectional events. While the amplitude of neural impulses is important, the temporal correlation of neuronal firing across these large networks is also considered to be critical. The most prominent dynamic principle underlying such correlation is reentry. Reentry is the large-scale recursive interaction of ongoing neural activity, both locally and in different areas across massively parallel reciprocal connections. It is most clearly seen within and between neural maps. The concepts of functional segregation, temporal correlation, and reentry are consistent with the units of selection existing at a much higher spatial scale than that of the functional components (such as the individual neurons) that play a major role in the received view. Indeed, in the radical view, the unit of selection necessary for perceptual categorization is considered to be a global mapping, a large-scale circuit made up both of cortical maps and subcortical structures whose operation is capable of yielding output that results in behavior.

In the radical view, the selectional system in the brain is capable of dynamic reconstruction of outputs under the constraints of value, selecting from a large number of degenerate possibilities. As such, this view rejects codes, representations, and explicit coded storage. Memory is nonrepresentational and is considered to reflect a dynamic capacity to recreate an act (or specifically to suppress one) under such constraints. The brain is not a computer, nor is the world an unambiguous piece of tape defining an effective procedure and constituting "symbolic information." Such a selectional brain system is endlessly more responsive and plastic then a coded system. In it the homunculus disappears, its role ceded to a self-organizing and necessarily complex biological system.

It is my belief that the radical view will become the received one within the next decade. Predictions are precarious, however, and we must remain open to the possibility that as a result of continuing discovery in neuroscience, an entirely new view of the brain will emerge.[29] One prediction nevertheless stands out as more secure than practically any other: the results of neuroscientific discovery will have enormous implications for our view of our position in nature. Establishing the proper picture of the brain will remain at the center of human concern.

ENDNOTES

[1]By the time of the French Revolution a picture was emerging of what was in the head, although almost a century would have to pass for the emergence of what we would accept as scientific data. An amusing indication of an earlier view can be found in *Le Rêve de d'Alembert* by Denis Diderot, in which d'Alembert's mistress, Mademoiselle de l'Espinasse, queries the physician, Dr. Bordeu, about the causes of d'Alembert's disturbed dreams:

BORDEU: Because it is a very different thing to have something wrong with the nerve-center from having it just in one of the nerves. The head can command the feet, but not the feet the head. The center can command one of the threads, but not the thread the center.

MADEMOISELLE DE L'ESPINASSE: And what is the difference, please? Why don't I think everywhere? It's a question I should have thought of earlier.

BORDEU: Because there is only one center of consciousness.

MADEMOISELLE DE L'ESPINASSE: That's very easy to say.

BORDEU: It can only be at one place, at the common center of all the sensations, where memory resides and comparisons are made. Each individual thread is only capable of registering a certain number of impressions, that is to say sensations one after the other, isolated and not remembered. But the center is sensitive to all of them; it is the register, it keeps them in mind or holds a sustained impression, and any animal is bound, from its embryonic stage, to relate itself to this center, attach its whole life to it, exist in it.

MADEMOISELLE DE L'ESPINASSE: Supposing my finger could remember.

BORDEU: Then your finger would be capable of thought.

MADEMOISELLE DE L'ESPINASSE: Well, what exactly is memory?

BORDEU: The property of the center, the specific sense of the center of the network, as sight is the property of the eye, and it is no more surprising that memory is not in the eye than that sight is not in the ear.

MADEMOISELLE DE L'ESPINASSE: Doctor, you are dodging my questions instead of answering them.

BORDEU: No, I'm not dodging anything. I'm telling you what I know, and I would be able to tell you more about it if I knew as much about the organization of the center of the network as I do about the threads, and if I had found it as easy to observe. But if I am not very strong on specific details I am good on general manifestations.

MADEMOISELLE DE L'ESPINASSE: And what might these be?

BORDEU: Reason, judgment, imagination, madness, imbecility, ferocity, instinct. . . .

BORDEU: And then there is force of habit which can get the better of people, such as the old man who still runs after women, or Voltaire still turning out tragedies.

(Here the doctor fell into a reverie, and MADEMOISELLE DE L'ESPINASSE said:) Doctor, you are dreaming.

BORDEU: Yes I was.

MADEMOISELLE DE L'ESPINASSE: What about?

BORDEU: Voltaire.

MADEMOISELLE DE L'ESPINASSE: What about him?

BORDEU: I was thinking of the way great men are made.

[2]Two reasonably elementary books for a lay reader are Gordon M. Shepard, *Neurobiology* (New York: Oxford University Press, 1983) and Gordon M. Shepard, *The Synaptic Organization of the Brain* (New York: Oxford University Press, 1990).

[3]Gerald M. Edelman and Vernon Mountcastle, *The Mindful Brain: Cortical Organization and the Group-Selective Theory of Higher Brain Function* (Cambridge, Mass.: MIT Press, 1978); Gerald M. Edelman, *Neural Darwinism: The Theory of Neuronal Group Selection* (New York: Basic Books, 1987); Jean-Pierre Changeux and Antoine Danchin, "Selective Stabilization of Developing Synapses as a Mechanism for the Specification of Neuronal Networks," *Nature* 264 (December 1976): 705–712.

[4]Shepard, *Neurobiology* and Shepard, *The Synaptic Organization of the Brain*; Eric Kandel, James H. Schwartz, and Thomas M. Jessell, eds., *Principles of Neural Science* (New York: Elsevier, 1991).

[5]Giulio Tononi, Olaf Sporns, and Gerald M. Edelman, "Reentry and the Problem of Integrating Multiple Cortical Areas: Simulation of Dynamic Integration in the Visual System," *Cerebral Cortex* 2 (July/August 1992): 310–335.

[6]Edelman, *Neural Darwinism*; Tononi, Sporns, and Edelman, "Reentry and the Problem of Integrating Multiple Cortical Areas"; Gerald M. Edelman, *The Remembered Present: A Biological Theory of Consciousness* (New York: Basic Books, 1989).

[7]Edelman, *Neural Darwinism*.

[8]Edelman and Mountcastle, *The Mindful Brain*.

[9]Ibid.; Gerald M. Edelman, "Through a Computer Darkly: Group Selection and Higher Brain Function," *Bulletin of the Academy of Arts and Sciences* 36 (October 1982): 20–49; Gerald M. Edelman, "Neural Darwinism: Selection and Reentrant Signaling in Higher Brain Function," *Neuron* 10 (February 1993): 115–125.

[10]Edelman and Mountcastle, *The Mindful Brain*; Edelman, *Neural Darwinism*; Edelman, *The Remembered Present*; Gerald M. Edelman, *Topobiology: An Introduction to Molecular Embryology* (New York: Basic Books, 1988).

[11]Tononi, Sporns, and Edelman, "Reentry and the Problem of Integrating Multiple Cortical Areas"; Olaf Sporns, Giulio Tononi, and Gerald M. Edelman, "Modeling Perceptual Grouping and Figure-Ground Segregation by Means of Active Reentrant Connections," *Proceedings of the National Academy of Science* 88 (January 1991): 129–133; George N. Reeke, Jr., Olaf Sporns, and Gerald M. Edelman, "Synthetic Neural Modeling: The 'Darwin' Series of Recognition Automata," *Proceedings of the Institute of Electrical and Elec-*

68   Gerald M. Edelman

tronics Engineers 78 (September 1990): 1498–1530; Karl Friston et al., "Value-Dependent Selection in the Brain: Simulation in a Synthetic Neural Model," Neuroscience 59 (March 1994): 229–243.

[12]Charles M. Gray and Wolf Singer, "Stimulus-Specific Neuronal Oscillations in Orientation Columns of Cat Visual Cortex," Proceedings of the National Academy of Science 86 (March 1989): 1698–1702 and Michael M. Merzenich et al., "Topographic Reorganization of Somatosensory Cortical Areas 3b and 1 in Adult Monkeys Following Restricted Deafferentation," Neuroscience 8 (January 1983): 33–55.

[13]Two nontechnical books are M. Mitchell Waldrop, Complexity: The Emerging Science at the Edge of Order and Chaos (New York: Simon & Schuster, 1992) and Roger Lewin, Complexity: Life at the Edge of Chaos (New York: Macmillan, 1992).

[14]Giulio Tononi, Olaf Sporns, and Gerald M. Edelman, "A Measure for Brain Complexity: Relating Functional Segregation and Integration in the Nervous System," Proceedings of the National Academy of Science 91 (May 1994): 5033–5037 and Giulio Tononi, Olaf Sporns, and Gerald M. Edelman, "A Complexity Measure for Selective Matching of Signals in the Brain," Proceedings of the National Academy of Science 93 (April 1996): 3422–3427.

[15]Tononi, Sporns, and Edelman, "A Measure for Brain Complexity."

[16]Tononi, Sporns, and Edelman, "A Complexity Measure for Selective Matching of Signals in the Brain."

[17]Tononi, Sporns, and Edelman, "Reentry and the Problem of Integrating Multiple Cortical Areas"; and Sporns, Tononi, and Edelman, "Modeling Perceptual Grouping and Figure-Ground Segregation."

[18]Daniel L. Schacter, Searching for Memory: The Brain, the Mind, and the Past (New York: Basic Books, 1996).

[19]Gerald M. Edelman, Bright Air, Brilliant Fire: On the Matter of the Mind (New York: Basic Books, 1992).

[20]Ibid.

[21]Edelman, The Remembered Present.

[22]Tononi, Sporns, and Edelman, "Reentry and the Problem of Integrating Multiple Cortical Areas"; Sporns, Tononi, and Edelman, "Modeling Perceptual Grouping and Figure-Ground Segregation."

[23]Edelman, The Remembered Present.

[24]Ibid.

[25]John R. Searle, The Mystery of Consciousness (London: Granta, 1997).

[26]In making this distinction, I am aware of the fact that certain scientists in neurobiology proper would agree, more or less, with the radical view. The fact remains, however, that many neurobiologists talk of the brain's carrying out of computations utilizing codes. In doing so, they do not account for the brain's enormous structural and dynamic variability. Moreover, cognitive psychologists, workers in artificial intelligence, and a good number of lin-

guists all hew to the notions I have bundled together as the received view. Naturally, there are as many variants in each position as there are individual scientists but the two positions are, I believe, usefully contraposed.

[27]George N. Reeke, Jr., and Gerald M. Edelman, "Real Brains and Artificial Intelligence," *Dædalus* 117 (1) (Winter 1988): 143–173.

[28]Edelman, *Neural Darwinism.*

[29]It is not possible to predict precisely what we will learn about the brain in the future. Consistent with my position here, I will put forth the conjecture that in carrying out a function such as speech, the interaction of the parts of the brain will not resemble those of an orderly machine but will more resemble a crazy quilt, the parts of which are connected in no uniformly systematic fashion. Evolution tinkers, it does not plan, and natural selection acts on functional outcomes of the individual organism, not on design drawings. I will also hazard a guess: as has been the case in other domains of science, once we know the facts and principles of brain function and structure more securely we will be able to imitate nature to a limited extent. At such a time, we will construct a conscious artifact. If we can give it the basis for language, we will be able to ask whether it categorizes or "carves nature at the joints" the way we do. The answer will perhaps provide as exciting or terrifying a prospect as hearing from intelligent life somewhere else in the galaxy. The ethical implications are obvious, if not easily resolvable. A more modest prediction is that we will build devices that incorporate what we have learned about brain functions and structures into new kinds of artifacts, mixing what we already know about computers with components capable of perceptual functions. This is already on its way.

As the twenty-first century opens, we can expect to witness the invention of even more sophisticated methods of brain imaging, as well as refinement of those already in use. With luck, scientists will eventually reach their ultimate goal of monitoring the activity of intact brains continuously and at the level of individual nerve fibers. In short, the mind as brain-at-work can be made visible.

Brain imaging and experimental brain surgery, together with analyses of localized brain trauma and endocrine and neurotransmitter mediation, have permitted a breakout from age-old subjective conceptions of mental activity. Researchers now speak confidently of a coming solution to the brain-mind problem.

Some students of the subject, however (including a few of the brain scientists themselves), consider that forecast overly optimistic. In their view, technical progress has been largely correlative and has contributed little to a deeper understanding of the conscious mind. They consider it the equivalent of mapping the communicative networks of a city, correlating its activity with ongoing social events, and then declaring the material basis of culture solved. Even if brain activity is mapped completely, they ask, where does that leave consciousness, and especially subjective experience? How to express joy in a summer rainbow with neurobiology? Perhaps these phenomena rise from undiscovered physicochemical phenomena or exist at a level of organization still beyond our comprehension. Or maybe, as a cosmic principle, the conscious mind is just too complicated and subtle ever to understand itself.

This view of the mind as *mysterium tremendum* is, in the opinion of most brain researchers, unjustifiably defeatist.

—Edward O. Wilson

"Consilience Among the Great Branches of Learning,"
from *Dædalus* Winter 1998,
"Science in Culture"

Semir Zeki

Art and the Brain

Les causeries sur l'art sont presque inutiles.[1]
　　　　　　　　　—*Paul Cézanne*

*More often than not, [people] expect a painting to
speak to them in terms other than visual, preferably
in words, whereas when a painting or a sculpture
needs to be supplemented and explained by words
it means either that it has not fulfilled its function
or that the public is deprived of vision.[2]*
　　　　　　　　　—*Naum Gabo*

I.

MUCH HAS BEEN WRITTEN ABOUT ART, but not in relation to
the visual brain—through which all art, whether in
conception or in execution or in appreciation, is expressed. A great deal, though perhaps not as much, has been
written about the visual brain, but little in relation to one of its
major products, art. It is therefore hardly surprising that the
connection between the functions of art and the functions of the
visual brain has not been made. The reason for this omission lies
in a conception of vision and the visual process that was largely
dictated by simple but powerful facts, derived from anatomy
and pathology. These facts spoke in favor of one conclusion to

---

*Semir Zeki is Professor of Neurobiology and Co-head of the Wellcome Department of
Cognitive Neurology at University College London.*

which neurologists were ineluctably driven, and that conclusion inhibited them, as well as art historians and critics, from asking the single most important question about vision that one can ask: Why do we see at all? It is the answer to that question that immediately reveals a parallel between the functions of art and the functions of the brain, and indeed ineluctably drives us to another conclusion—that the overall function of art is an extension of the function of the brain. In that definition are the germs of a theory of art that has solid biological foundations and that unites the views of modern neurobiologists with those of Plato, Michaelangelo, Mondrian, Cézanne, Matisse, and many other artists.

The concept of the functions of the visual brain inherited by modern neurobiologists was based on facts derived between 1860 and 1970. Chief among these was the demonstration by the Swedish neuropathologist Salomon Henschen and his successors in Japan and England that the retina of the eye is not diffusely connected to the whole brain, or even to half the brain, but only to a well-defined and circumscribed part of the cerebral cortex. First called the visuo-sensory cortex and later the primary visual cortex (area V1), it therefore constituted "the only entering place of the visual radiation into the organ of psyche."[3] This capital discovery led to a prolonged battle between its proponents and its opponents, who thought of it as "une localisation à outrance";[4] they had conceived of the visual input to the brain as being much more extensive and as including large parts of the cerebral cortex that were known to have other functions, a notion more in keeping with the doctrine of the French physiologist Pierre Flourens. The predecessor of the American psychologist Karl Lashley, Flourens had imagined that each and every part of the cortex is involved in every one of its activities. It was not until early this century that the issue of a single visual area located in an anatomically and histologically defined part of the cortex was settled in favor of the localizationists.[5] There was much else to promote the idea of V1 as the "sole" visual center. It had a mature appearance at birth, as if ready to "receive" the visual "impressions formed on the retina,"[6] whereas the cortex surrounding it matured at different stages after birth, as if the maturation depended upon the acquisition of experience; this made of the latter higher cognitive centers, the *Cogitatzionzentren,* whose function was to interpret the visual image received by V1, or so neurologists imagined. As well, lesions in V1 lead to blindness, the position and extent of which is in direct proportion to the position and size of the lesion; by contrast, lesions in the surrounding cortex resulted in vague visual syndromes, referred to first as mind blindness (*Seelenblindheit*) and then as agnosia, following the term introduced by Freud. Together, these facts conferred the sovereign capacity of "seeing" on V1, leading neurologists to conceive of it as the "cortical retina," the cerebral organ that receives the visual images "impressed" upon the retina, as on a photographic plate—an analogy commonly made. Seeing was therefore a passive process, while understanding what was seen was an active one; this notion divided seeing from understanding and assigned a separate cortical seat to each.

This concept left little room for the fundamental question of why we see. Instead, seeing was accepted as a given. Asked the question today, few would suppose that it is so we can appreciate works of art; most would give answers that are specific, though related in general to the survival of the species. The most general of these answers would include all the specific ones and define the function of seeing as *the acquisition of knowledge about the world.*[7] There are, of course, other ways of obtaining that knowledge—through the sense of touch or smell or audition, for example—but vision happens to be the most efficient way of obtaining it, and there are some kinds of knowledge, such as the color of a surface or the expression on a face, that can only be obtained through vision.

It takes but a moment's thought to realize that obtaining that knowledge is no easy matter. The brain is only interested in obtaining knowledge about those permanent, essential, or characteristic properties of objects and surfaces that allow it to categorize them. But the information reaching the brain from these surfaces and objects is in continual flux. A face may be categorized as a sad one, thus giving the brain knowledge about a person in spite of the continual changes in individual features, viewing angle, or even the identity of the face viewed; or the destination of an object may have to be decided by its direction of motion, regardless of its speed or distance. An object may have to be categorized according to color, as when judging the state of ripeness of an edible fruit. But the wavelength composition of the light reflected from an object is never constant; it changes continually, depending upon the time of day, without entailing a substantial shift in its color. The ability of the brain to assign a constant color to a surface or a constant form to an object is generally referred to as color or object constancy. But perceptual constancy is a much wider phenomenon. It also applies, for example, to faces that are recognizable when viewed from different angles and regardless of the expression worn. There is also what I shall call situational constancy, when the brain is able to categorize an event or a situation as festive or sad and so on, regardless of the particular event. There is even a narrative constancy when, for example, the brain is able to identify a scene as the "Descent from the Cross," regardless of variations in the detail or the style of the painting. The brain, in each case, extracts from the continually changing information that reaches it only what is necessary for it to identify the characteristic properties of what it views; it has to extract constant features in order to be able to obtain knowledge about them and to categorize them. Vision, in brief, is an active process that depends as much upon the operations of the brain as upon the external, physical environment; the brain must discount much of the information reaching it, select only what is necessary in order to obtain knowledge about the visual world, and compare the selected information with its stored record of all that it has seen. A modern neurobiologist should approve heartily of Matisse's statement that "Voire, c'est déjà une operation créatrice, qui exige un effort."[8]

How the brain achieves this remarkable feat remains a puzzle; indeed, the question has only been seriously addressed in the last thirty years, which have witnessed a prolific output of work on the visual brain. Among the chief discoveries is that it is composed of many different visual areas that surround V1.[9] Each group of areas is specialized to process a particular attribute of the visual environment by virtue of the specialized signals that each receives from V1.[10] Cells specialized for a given attribute such as motion or color are grouped together in anatomically identifiable compartments within V1, and different compartments connect with different visual areas outside V1, thus conferring their specializations on the relevant areas.[11] V1 acts much like a post office, distributing different signals to different destinations; it is but the first, though essential, stage in an elaborate machinery designed to extract the essential information from the visual world. What we now call the visual brain is actually V1 plus the specialized visual areas with which it connects, directly and indirectly. We therefore speak of parallel systems devoted to processing simultaneously different attributes of the visual world, a system comprising the specialized cells in V1 plus the specialized areas to which these cells project. Vision, in brief, is modular. The reasons for evolving a strategy to process in parallel the different attributes of the visual world have been debated, but it seems plausible to suppose that they are rooted in the need to discount different kinds of information when acquiring knowledge about different attributes.[12] With color, it is the precise wavelength composition of the light reflected from a surface that has to be discounted, whereas with size it is the precise viewing distance, and with form, the viewing angle.

Recent evidence has shown that the processing systems are also perceptual systems in that activity in each can result in a percept without reference to the other systems; each processing-perceptual system terminates its perceptual task and reaches its perceptual endpoint at a slightly different time than the others, thus leading to a perceptual asynchrony in vision—color is seen before form, which is seen before motion, with the advantage of color over motion being on the order of 60–100 ms.[13] Thus visual perception is also modular. In summary, the visual brain is characterized by a set of parallel processing-perceptual systems and a temporal hierarchy in visual perception.[14]

These findings lead me to propose that there is also a modularity, a functional specialization, in visual aesthetics. When area V4, the color center, is damaged, the consequence is an inability to see the world in color.[15] But other attributes of the visual scene are perceived normally. When area V5, the motion center, is damaged, the consequence is an inability to see objects when in motion, although other attributes are seen normally. Damage to a region close to V4 leads to a syndrome characterized by an inability to see familiar faces. There are other specific syndromes—for example, the inability to recognize certain categories of objects—and neurology is continually uncovering new syndromes of selective visual loss. I do not mean, of course, to imply that the aesthetics of color are due solely to the activity in V4 or the aesthetics of kinetic art are due solely to the activity in V5; I am suggesting only that the perception of color and of motion is not possible without the presence and healthy functioning of these areas. It does little good to ask a patient with a V4 lesion to appreciate the complexities of fauvist art or a patient with a V5 lesion to view the works of Tinguely. These are aesthetic experiences of which such patients are not capable.

II.

The definition that I have given above of the function of the visual brain—a search for constancies with the aim of obtaining knowledge about the world—is applicable with equal vigor to the function of art. I shall thus define the general function of art as a search for the constant, lasting, essential, and enduring features of objects, surfaces, faces, situations, and so on, which allows us not only to acquire knowledge about the particular object, or face, or condition represented on the canvas but to generalize, based on that, about many other objects and thus acquire knowledge about a wide category of objects or faces. In this process, the artist must also be selective and invest his work with attributes that are essential, discarding much that is superfluous. It follows that one of the functions of art is an extension of the major function of the visual brain. Indeed, philosophers and artists often spoke about art in terms that are extremely similar to the language that a modern neurobiologist of vision would use, except that he would substitute the word "brain" for the word "artist." It is striking, for example, to compare Herman von Helmholtz's statement about "discounting the illuminant," in which a colored surface is viewed (in order to assign a constant color to a surface), with the statement of Albert Gleizes and Jean Metzinger in their book *Cubism*. Discussing Gustave Courbet, they wrote, "Unaware of the fact that in order to display a true relation we must be ready to sacrifice a thousand apparent truths, he accepted, without the slightest intellectual control, all that his retina presented to him. He did not suspect that the visible world can become the real world only by the operation of the intellect."[16] I interpret "intellect" to mean the brain or, better still, the cerebral cortex. In order to represent the real world, the brain (or the artist) must discount ("sacrifice") a great deal of the information reaching it (or him), information that is not essential to its (or his) aim of representing the true character of objects.

It is for this reason that I hold the somewhat unusual view that artists are neurologists, studying the brain with techniques that are unique to them and reaching interesting but unspecified conclusions about the organization of the brain. Or, rather, that they are exploiting the characteristics of the parallel processing-perceptual systems of the brain to create their works, sometimes even restricting themselves largely or wholly to one system, as in kinetic art. These conclusions are on canvas and are communicated and understood through the visual medium, without the necessity of using words. This may surprise them since most of them, naturally enough, know nothing about the brain, and a good many still hold the common but erroneous belief that one sees with the eye rather than with the cerebral cortex. Their language, as well as the language of those who write about art, betrays this view. But however erroneous their views about the seeing organ or the role of the visual brain may be, it is sufficient to glance at their writings to realize the extent to which they have defined the function of art in a way that a modern neurobiologist would not only understand but feel very sympathetic to. Thus Henri Matisse once said, "Underlying this succession of moments which constitutes the superficial existence of things and beings, and which is continually modifying and transforming them, one can search for a truer, more essential character, which the artist will seize so that he may give to reality a more lasting interpretation."[17] Essentially, this is what the brain does continually—seizing from the constantly changing information reaching it the more essential one, distilling from the successive views the essential character of objects and situations. Similar statements abound, and it is sufficient to give just one more example. Jacques Rivière, the art critic, wrote: "The true purpose of painting is to represent objects as they really are, that is to say, differently from the way we see them. It tends always to give us their sensible essence, their presence, this is why the image it forms does not resemble their appearance . . ." because the appearance changes from moment to moment.[18] A neurologist could hardly have improved on that statement in describing the functions of the visual brain. He might say that the function of the brain is to represent objects as they really are, that is to say, differently from the way we see them from moment to moment if we were to take into account solely the effect that they produce on the retina.

To summarize, therefore, both the brain and one of its products, art, have the task of, in the words of artists themselves, depicting objects as they are. And both face the problem of how to distill from the ever changing information in the visual world only that which is important in order to represent the permanent, essential characteristics of objects. Indeed, this was almost the basis of Kant's philosophy of aesthetics, to represent perfection; but perfection implies immutability, and hence arises the problem of depicting perfection in an ever changing world. I shall therefore define the function of art as being a search for constancies, which is also one of the most fundamental functions of the brain. The function of art is therefore an extension of the function of the brain—the seeking of knowledge in an ever-changing world.

III.

Plato was among the most prominent of those who lamented the poverty of art. Without saying so, and indeed without ever referring to the brain, he implicitly compared the limitations of art to the infinite capacities of the brain. His most explicit statement in this regard occurs in Book X of *The Republic*, where he dismisses painting as a mimetic art, one that could represent only one aspect of a particular example of a more general category of an object. To him there was the general ideal of a given form, which was the embodiment of all the examples of that form; then there was a particular form that was but one example of the more general, "universal" form; and finally, there was painting, which captured but one facet, one image, of one particular form. "The Greeks," Sir Herbert Read tells us,[19] "with more reason, regarded the ideal as the real, and representational art as merely an imitation of an imitation of the real."[20]

Plato's contempt for painting was really linked to his theory of forms and ideals. The example he gives in Book X is that of a couch. To him, there is only one real couch, the one created by God; this is the idea of a couch, and it has a universal existence. One can therefore obtain real knowledge only about this one ideal couch. Of particular couches—as made by a craftsman (δημιουργοζ), or represented in a single view in a painting, or reflected in a mirror—there can only be an opinion, and an unverifiable one at that.[21] Put in mathematical terms, we can only obtain real and reliable knowledge about ideal circles, triangles, and straight lines. Viewing painted circles and straight lines without reference to the Ideal leads only to a superficial impression and an opinion, which may turn out to be true or false. Plato implied that, at least to get nearer to the Ideal, painting should change direction in order to represent as many facets of an object or situation as possible, since this would give more knowledge about the object. What Plato only implied, Schopenhauer made explicit many centuries later when he wrote that painting should strive "to obtain knowledge of an object, not as a particular thing but as Platonic Ideal, that is, the enduring form of this whole species of things," a statement that a modern neurobiologist could easily accommodate in describing the functions of the visual brain.[22] Indeed, to a neurobiologist, a brain that is not able to do this is a sick, pathological brain. Painting, in other words, should be the representation of the constant elements, the essentials, that would give knowledge of all couches; it should represent constancies. As John Constable put it in his *Discourses*: "The whole beauty and grandeur of Art consists . . . in being able to get above all singular forms, local customs, particularities of every kind . . . [The painter] makes out an abstract idea of their forms more perfect than any one original," the "abstract idea" presumably being Constable's term for the Platonic Ideal.[23]

There is something unsatisfactory about the Platonic Ideal from a neurobiological point of view, because the Ideal has an existence that is external to the brain and without reference to it; we can only have an opinion of that which we perceive, "whereas knowledge is of a super-sensible eternal world."[24] Implicitly more dependent upon brain function, and thus more acceptable neurobiologically, are the views of Kant and Hegel. Their views exalt art, which is seen as being able to represent reality better than the "ephemera of sense data," since the latter changes from moment to moment. Hegel deals with the Idea that is derived from the Concept. In a painting, the brain, having "accumulated a treasure," can "now freely disgorge [it] in a simple manner without the far-flung conditions and arrangements of the real world." By this process of "disgorging," and thus of externalizing and concretizing, the Concept becomes the Idea. The Idea, then, is merely the external representation of the Concept that is in the brain, the Concept that it has derived from ephemeral sense data. It is, in fact, the product of the artist. Art, including painting, therefore, "furnishes us with the things themselves, but out of the inner life of the mind"; through art, "instead of all the dimensions requisite for appearance in nature, we have just a surface, and yet we get the same impression that reality affords."[25] It is through this translation of the Concept into the Idea that Dutch painting, for example, "has recreated . . . the existent and *fleeting appearance of nature* as something generated afresh by man."[26]

Although the views of Plato and Hegel may appear antipodean, the difference between the two is in fact neurobiologically irrelevant if we try to give a neurobiological definition of the Platonic Ideal and the Hegelian Concept. The first step in such a definition, relevant to Plato's views but less so to Hegel's, is a neurobiological doctrine: forms do not have an existence without a brain. This may seem an audacious statement to make, but it is supported by numerous clinical and physiological studies that have shown that individuals who are born blind and to whom vision is later restored find it very difficult, if not impossible, to learn to see even a few forms, and these they soon forget. The question that the learned Mr. Molyneux asked in John Locke's *Essay Concerning Human Understanding*—whether a man born blind and who had learned to distinguish between forms by touch alone would be able to distinguish them by sight alone when vision is restored to him—has been answered negatively many times by clinical studies.[27] Physiological studies, particularly those of David Hubel and Torsten Wiesel, have shown that even when the genetically determined visual apparatus is intact at birth, the organism must be exposed to visual stimuli after birth, after which visual education becomes much less important.[28] There is, in other words, a critical period for vision, just as there appears to be for emotional development.[29] Artists have often wished that they could see and paint the world as a child does—for the first time, innocently, without what they suppose to be the prejudice of the developed and possibly even corrupted influence of a brain that has knowledge of the world. Picasso admired the art of children and Matisse wanted to paint like them, as does Balthus;[30] Monet wished that he could have been born blind, with vision restored to him later in life so that he could see pure form "without knowing what the objects were that he saw before him."[31] They are all yearning for something that is physiologically almost impossible. The visual apprenticeship of children occurs at a very early age, before two, and begins immediately after birth, long before the motor apparatus has developed sufficiently to be able to execute a painting. In its conceptual immaturity and technical simplicity, the art of a four-year-old child may be touching and even exciting, but it is the art of a visual brain that is already highly developed and that has acquired much knowledge about the world. The innocence that artists yearn for is, in terms of the brain, a myth.

If neurologically no forms, ideal or otherwise, exist without a brain that is properly nourished, how can we define the Platonic Ideal and the Hegelian Concept in neurological terms? I would propose that both can be equated with the brain's stored memory record of all the views of all the objects that it has seen, from which it has formed a Concept or an Ideal of these objects such that a single view of an object makes it possible for the brain to categorize that object. Indeed, in Plato's system, we can only recognize and categorize objects of which our immortal souls have seen examples constructed by δημιουργο (see, for example, Plato's *Meno*). In this sense, therefore, the Platonic system acknowledges the importance of a stored record, though without making reference to the brain. The recognition that we can only categorize objects that we have already seen (and therefore have a general representation of) nevertheless constitutes a far-reaching insight and brings Plato's position close to a modern neurobiological one. Neurobiology would have to depart from the Platonic system in saying not only that this general representation is built by the brain but also that there can be no Ideals without the brain.

We know a little, but not much, about the brain's stored visual memory system for objects. We know that it must involve a region of the brain known as the inferior convolution of the temporal lobes, because damage here causes severe problems in object recognition. Although very much in their infancy, recent physiological studies have started to give us some insights into the more detailed physiological mechanisms involved.[32] When a monkey, an animal that is close to man, is exposed to different views of objects that it has never encountered before (objects generated on a television screen), recording from single cells in its inferior temporal cortex can show how they respond when these same objects are subsequently shown on the screen again. Most cells respond to one view only, and their response declines as the object is rotated in such a way as to present increasingly less-familiar views. A minority of cells respond to only two views, but only a very small proportion, amounting to less than 1 percent, respond in a view-invariant manner. Whether they respond to one or more views, the actual size of the stimuli or the precise position in the field of view in which they appear make little difference to the responses of the cell. On the other hand, no cells have ever been found that are responsive to views with which the animal has not been familiarized; hence, exposure to the stimulus is necessary, from which it follows that the cells may be plastic enough to be "tuned" to one or more views of an object. In summary, many cells, each one responsive to one view only, may be involved during recognition of an object, with the whole group acting as an ensemble. But the presence of that small 1 percent of cells that respond in a view-invariant manner suggests also that form constancy may be the function of a specialized groups of cells, since 1 percent represents an enormous number in absolute terms.

When undertaking their work, artists generally are concerned not with philosophical views but rather with achieving desired effects on canvas—by experimenting, by "sacrificing a thousand apparent truths" and distilling the essence of their visual experience. We are told, for example, that Cézanne's work is "a painted epistemology" (*Erkenntnis Kritik*), since Cézanne supposedly shared Kant's ideology.[33] But Cézanne, in particular, put paid to all these empty speculations even before they were made when he said that "les causeries sur l'art sont presque inutiles."[34] I agree with Kahnweiler when he states, "J'insiste, en passant, sur le fait qu'aucun de ces peintres . . . n'avait de culture philosophique, et que les rapprochments possibles—avec Locke et Kant surtout—d'une telle attitude leur étaient inconnus, *leur classement étant, d'ailleurs, instinctif plus que raisonné.*"[35] The preoccupation of artists has instead been less exalted and more similar to the physiological experiments described earlier: exposing themselves to as many views of their subject as possible and thus obtaining a brain record from which they can distill on canvas the best combination. If, in executing his work, the artist is indifferent to these polar views—Plato on the one hand, and Hegel and Kant on the other—so should the neurobiologist be, if he accepts my equation of the Platonic Ideal and the Hegelian Concept with the brain's stored record of what it has seen. Whether art succeeds in presenting the real truth, the essentials, or whether it is the only means of getting to that truth in the face of constantly changing and ephemeral sense data, the opposing views are at least united in suggesting that there is (Hegel) or that there should be (Plato and Schopenhauer) a strong relationship between painting and the search for essentials. And my equation of both the Hegelian Concept and the Platonic Ideal with the brain's stored record means that the difference between the two, from a neurological point of view, is insignificant. There have been artists who have, again without reference to the brain or its stored record, tried deliberately and with much success to contradict the stored memory record of the brain. Many of the works of René Magritte go against everything that the brain has seen, learned, and stored in its memory. There is no Platonic Ideal or Hegelian Concept here because the brain has no representation of such bizarre scenes. It is an act of the imagination that fascinates the brain, which tries to make sense of a scene that goes against all its experience and for which it can find no solution.

IV.

To a neurobiologist viewing the art scene without being involved in it, it seems to be cubism that most explicitly, without acknowledging it or perhaps being even aware of it, set out to answer that deep paradox between reality and appearance alluded to by Plato—although this is my interpretation, not that of cubists. Cubism, the most radical departure in Western art since Paolo Uccello and Piero della Francesca introduced perspective into painting, "was a sort of analysis," a static representation of the result of "moving around an object to seize several successive appearances, which, fused in a single image, reconstitute it in time."[36] The aims of cubist painting, which was an attempt "to discover less unstable elements in the objects to be represented," were well stated by the French critic Jacques Rivière, and they read as if they were an account of the aims of the brain.[37] Rivière wrote that "The cubists are destined . . . to give back to painting its true aims, which is to reproduce . . . objects as they are." But to achieve this, "lighting must be eliminated" because "it is the sign of a particular instant. . . . If, therefore, the plastic image is to reveal the essence and permanence of things, it must be free of lighting effects. . . . It can therefore be said that lighting prevents things from *appearing as they are*. . . . Sight is a successive sense; we have to combine many of its perceptions before we can know a single object well. But the painted image is fixed. . . ." Perspective must also be eliminated because it "is as accidental a thing as lighting. It is the sign . . . of a particular position in space. It indicates not the situation of objects but the situation of a spectator. . . . Perspective is also the sign of an instant, of the instant when a certain man is at a certain point."[38] That statement is one that a modern neurobiologist would applaud, for the brain likewise never sees the objects and surfaces that make up the visual world around us from a single point or in a standard lighting condition. Instead, objects are viewed at different distances, from different angles and in different lighting conditions, yet they maintain their identity.

The solution that cubism brought to this problem was to try to mimic the functions of the brain, though with far less success. The precursor of cubist painting is generally agreed to be Picasso's *Les Demoiselles d'Avignon,* a forceful painting about which a great deal has been written, much of it neurologically and visually uninformative. What is especially interesting visually is the ambiguity in the figures, especially the one seated to the bottom right; she could be facing us or facing sideways. This ambiguity was much exploited by both Picasso and the cofounder of cubism, Georges Braque. The elimination of the point of view became a prominent feature of many of Picasso's portraits, so that the subject could seemingly be facing one direction or another. In later representative paintings such as *The Violin Player,* Picasso introduced so many different points of view that the painting ceased to be recognizable to the human brain, the final result being only recognizable as a violin player through its title. A brain ignorant of that title can hardly construe this as a violin player. The brain regularly views objects and people from different angles and is able to integrate these different views in an orderly way, allowing it to recognize and obtain knowledge about what it is viewing. The attempt by cubism to mimic what the brain does, to create a perceptual constancy for objects regardless of viewing angle, was, in the neurobiological sense, a failure—a heroic failure perhaps, but a failure nevertheless. My neurobiological interpretation is that it is indeed because of this failure that cubism changed course and entered its later, synthetic phase; it is certain that Mondrian saw it that way, for he abandoned cubism and accused it of "not accept[ing] the logical consequences of its own discoveries [and] developing abstraction towards its ultimate goal, the expression of pure reality."[39] In the synthetic phase, Kasimir Malevich tells us, "objective nature is merely *the starting point—the motivation—for the creation of new forms,* so that the objects themselves can scarcely, if at all, be recognized in the pictures."[40] But the new forms that synthetic cubism created were ultimately derived from the forms in nature that the artist was exposed to, and perhaps the best proof of this is found in the objective titles given to the paintings. It is difficult for the brain of a spectator to decipher what many of the creations of synthetic cubism represent. It was probably also difficult for Picasso himself, which is presumably one reason why he used objective and recognizable titles to describe his paintings. Nilsen Lauvrik, who was hostile to cubism, described *Woman with a Mustard Pot* as "one of the most engaging puzzles of a very puzzling art. This is sharply emphasized by the delight and pride of every spectator who is successful in solving the puzzle by finding in these enigmatic charts some sort of a tangible, pictorial justification of the title appended thereto. . . . The discovery of the 'mustard pot' would scarcely have been possible without the happy cooperation of the title with the spectator's previous knowledge of the actual appearance of a mustard pot."[41]

V.

From the neurobiological point of view, representational art was a good deal more successful in meeting the brain's incessant demands for constancy. Here I will consider neurologically the work of two different artists, Vermeer and Michelangelo, both of whom—unknowingly and in different ways—satisfied this demand far better than the product of the heroic but neurobiologically flawed experiments of the cubists.

A great deal has been written about Vermeer, "un artiste à jamais inconnu," as Proust astutely called him.[42] His technical brilliance, his use of perspective and rich chromatism are all common knowledge. But in viewing a painting such as *Man and Woman at the Virginal* (displayed at Buckingham Palace), it is not these features that attract and move the ordinary viewer. Paul Claudel, among others, has commented on the banality of Vermeer's subjects: an interior, a maid pouring milk, a girl weighing gold, another reading a letter, a music lesson—all daily events seemingly without special significance.[43] But there is, in Claudel's words, something "eerie, uncanny" about them.[44] In a good many of his paintings, the viewer is invited to look inside, as if through a keyhole, but not to enter.[45] He is a voyeur, peering into the private moments of private, unknown, individuals; what they are doing, or saying, or thinking is a mystery. It is this aspect of Vermeer, I believe, that has the immediate power to attract and provoke, and his technical virtuosity is used in the service of that psychological power, not as an end in itself, unsurpassed though it may be.

Where does this psychological power come from and what, in any case, do we mean by psychological power? A painting like *Man and Woman at the Virginal* derives its grandeur, I believe, from the way in which its technical virtuosity is used to generate ambiguity. I use the term ambiguity here to mean its ability to represent simultaneously, on the same canvas, not one but several truths, each as valid as the others.[46] These several truths revolve around the relationship between the man and the woman. There is no denying that there is some relationship between them. But is he her husband, her lover, a suitor, or a friend? Did he actually enjoy the playing or does he think that she can do better? Is the harpsichord really being used or is she merely playing a few notes while concentrating on something else, perhaps something he told her, announcing a separation or a reconciliation, or maybe something a good deal more banal? All these scenarios have equal validity in the painting, which can thus satisfy several "ideals" simultaneously. Through its stored memory of similar past events, the brain can recognize in this work the ideal representation of many situations and can categorize the scene represented as happy or sad. This gives ambiguity, a characteristic of all great art, a different, and neurological, definition: not the vagueness or uncertainty found in dictionaries, but, on the contrary, certainty—the certainty of many different and essential conditions, each equal to the others, all expressed in a single painting, profound in its faithful representation of so much.

The Vermeer painting satisfies Schopenhauer's wish that a painting should "obtain knowledge of an object, not as particular thing but as Platonic Ideal, that is to say, the enduring form of this whole species of thing."[47] In any of a number of situations, the scene depicted is what one might actually expect. There is a constancy about it, which makes it independent of the precise situation and applicable to many. The painting is indeed "a vision of two distant people 'alone together' in a space moved by forces beyond the ken of either," a scenario effectively exploited by Michaelangelo Antonioni in some of his films, and most notably in *l'Avventura* and *l'Eclisse*, where once again the viewer becomes imaginatively involved in trying to guess the thoughts of the protagonists.[48] Though it may come as a surprise, there is in this respect and in terms of the brain a certain neurobiological similarity between the paintings of Vermeer and those of the cubists, especially the later variety that cultivated an ambiguity, in the sense that I have used the term. Writing of cubism, Gleizes and Metzinger tell us that "certain forms should remain implicit, so that the mind of the spectator is the chosen place of their concrete birth."[49] There could be no more admirable description of the work of Vermeer, where very nearly all is implicit. As with forms and objects in cubist art, the brain of the spectator is the chosen birthplace of many situations in Vermeer's paintings, each one of equal validity as the others. The true solution remains "à jamais inconnu," because there is no true solution, there is no correct answer. It is therefore a painting for many conditions.

Situational constancy is a subject that neurology has not yet studied; indeed, the problem itself has not been addressed. We have hardly begun to understand the simpler kinds of constancy, such as form or color, and it is not surprising that neurologists should not have even thought of studying so complex a subject. I would guess from the kind of physiological experiment described above that, in broad outline, exposure of an individual to a few situations would be sufficient to extract the elements that would be common to all similar occasions. But what brain mechanisms are involved remains a mystery today.

It was perhaps the masters of Cycladic art in the sixth century B.C. who understood earlier than most that the brain must be the place of birth of implicit forms. They created works that emphasized a few organs—the lips or the nose, for example—and left it to the imagination of the beholder to complete the form. Michaelangelo achieved much the same effect by leaving many of his sculptures unfinished. Why he did so remains a question of debate; my interpretation is that this was one solution to the problem of representing the many facets of spiritual beauty and divine love—it was too great a task even for the mighty Michaelangelo. We know that he usually refused to execute portraits, believing that he could not represent all the beauty that his brain had formed a Concept of. Two exceptions are his portraits of Andrea Quaratesi and of Tomasso de' Cavalieri, the young nobleman who had overwhelmed him with his beauty and had come to dominate his emotional life in his later years, unleashing a furious creative energy of great brilliance. But the difficulty of portraying physical beauty was nothing compared to that of depicting spiritual beauty and divine love. As a devout Catholic, Michaelangelo found that love in the life of Jesus, and particularly in the last moments on the Cross and after the descent from it, the subject of several of his sculptures. By leaving many of his sculptures unfinished—for example, the *Rondanini Piéta* and the *San Matteo*—Michaelangelo invites the spectator to be imaginatively involved, and the spectator's view will fit many of the Concepts, the stored representations, in his brain. The forms that the unfinished work merely suggests become concretely realized only in the spectator's brain. There is an ambiguity here too, and therefore a constancy about these unfinished works, but the ambiguity is reached by a different route. Perhaps the most definitive hint at what Michaelangelo intended is found in his *Rime* or *Sonnets,* where, next to his works, he best expounds his views on art and beauty. In one, dedicated to Vittoria Colonna, Marchesa di Pescara, he wrote:

> The greatest artists have no thought to show that
> Which the marble in its superfluous shell does not contain
> To break the marble spell is all that the hand
> That serves the brain can do[50]

VI.

The Alexandrian Neo-Platonist Plotinus, with whose writings Michaelangelo was surely acquainted, remarked, "The form is in the sculptor long before it ever enters the stone," a biological truth that enables the sculptor to fashion his work and the spectator to appreciate it.[51] But if the form is in the artist (and the spectator), maybe neither need the forms in the outside world at all. That was the real starting point for the work of the Russian suprematist Kasimir Malevich, a neurobiologically interesting artist, who wrote that "art wants nothing further to do with the objective world as such." The use of the word "further" here gives biological credibility to Malevich's doctrine because, as discussed above, the brain needs to be visually nourished at critical periods after birth so as not to remain almost indefinitely blind. So the nonobjective sensation and art of which Malevich speaks is really the introspective art of a brain that is well acquainted with the objective world; it has already selected all the information necessary to be able to identify and categorize objects. And true to its aims of being a search for essentials and constants, we find that as art developed more and more in the modern era, much of it became better and better tailored to the physiology of the parallel processing-perceptual systems and the visual areas that we have only recently discovered, and specifically to the physiology of single cells in them. The physiology of these areas is itself tailored to extract the essential information in the visual environment—there exists an *Einfühlung,* that untranslatable term signifying a link between the "preexistent" forms within the individual and the forms in the outside world that are reflected back, the "art de peindre des ensembles nouveaux empruntés non à la réalité visuelle, mais à celle que suggèrent à l'artiste l'instinct et l'intuition," as Guillaume Apollinaire said of cubism.[52]

Physiologically, the *Einfühlung* is expressed in what I have called the art of the receptive field and I shall give but two brief examples of it here.[53] The receptive field is one of the most important concepts to emerge from sensory physiology in the past fifty years. It refers to the part of the body (in the case of the visual system, the part of the retina or its projection into the visual field) that, when stimulated, results in a reaction from the cell, specifically, an increase or decrease in its resting electrical discharge rate. To be able to activate a cell in the visual brain, one must not only stimulate in the correct place (i.e., stimulate the receptive field) but also stimulate the receptive field with the correct visual stimulus, because cells in the visual brain are remarkably fussy about the kind of visual stimulus to which they will respond. The art of the receptive field may thus be defined as that art whose characteristic components resemble the characteristics of the receptive fields of cells in the visual brain and which can therefore be used to activate such cells.

One group of cells, discovered by David Hubel and Torsten Wiesel in 1959, will only respond to lines of particular orienta-, tion, since the orientational preferences of different cells are different and each responds increasingly more grudgingly as one departs from the preferred orientation until the response disappears at the orthogonal orientation. Such cells are a prominent feature of area V1 and some other areas surrounding it, notably V3, but they are also found in other areas. They are usually considered to be the physiological "building blocks" of form perception, though how one moves from such cells to the creation of forms remains unknown. It is interesting that the line is among the most prominent features of the "nonobjective" art of Malevich and his successors. Lines are the predominant and sometimes only feature in the paintings of artists as diverse as Olga Rozanova, Barnett Newman, Robert Motherwell, Ellsworth Kelly, Gene Davis, Robert Mangold, and Ad Reinhardt, to mention but a few. Together with the rectangle and the circle, they were considered by Malevich to be the most elemental aspect of the nonobjective world. Mondrian, too, came to emphasize lines but reached that conclusion from an intellectually (though perhaps not physiologically) different route. Art, he believed, "shows us that *there are also constant truths* concerning forms," and it was the aim of objective art, as he saw it, to reduce all complex forms to one or a few universal forms, the constant elements that would be the constituent of all forms. Thus, as he said, to "discover consciously or unconsciously the fundamental laws hidden in reality" and "to create pure reality plastically it is necessary to reduce natural forms to the *constant elements*."[54] He sought, in other words, the Platonic Ideal for form (though he did not describe it in these terms). This search led to the vertical and horizontal lines, or so he believed. These "exist everywhere and dominate everything." Moreover, the straight line "is a stronger and more profound expression than the curve" because "all curvature resolves into the straight, no place remains for the curved."[55] He wrote, "Among the different forms, we may consider those as being neutral which have neither the complexity nor the particularities possessed by natural forms or abstract forms in general."[56]

This emphasis on line in many of the more modern and abstract works of art, in all probability, is derived not from a
profound knowledge of geometry but simply from the experimentation of artists to reduce the complexity of forms into their
essentials or, to put it in neurological terms, to try and find out
what the essence of form as represented in the brain may be.
This is my interpretation, not that of artists, but I cannot see
that it is any less valid than other interpretations. Kahnwiler, the
art dealer, tells us, "C'est uniquement l'apparition, chez les
cubistes, de lignes droites . . . qui a fait croire a une geométrie
dont il n'y a, en réalité, nulle trace. Ces droites . . . reflets de la
base meme, de l'à priori, de toute perception visuelle humaine,
se retrouvent, en fait, dans toute oeuvre d'art plastique des que
le souci d'imitation a disparu."[57] This is as explicit a statement
as any—coming from one who, if not an artist himself, was at
least well acquainted with artists and their work—that the artist
is trying to represent the essentials of form as constituted in his
visual perception, which I take to mean the brain. Gleizes and
Metzinger, both artists, emphasized the straight lines and the
relationship that they have to each other, as did Mondrian. They
wrote, "The diversity of the relations of line to line must be
indefinite; on this condition it incorporates quality, the incom-
mensurable sum of the affinities perceived between that which
we discern and that which pre-exists within us."[58] Once again, I
interpret "that which pre-exists within us" to mean that which
is in our brains. Although Gleizes and Metzinger are discussing
here the relations between lines, it is nevertheless lines that they
have chosen to emphasize. Equally interesting are the speculations of Mécislas Golberg, a man said to have had a powerful
influence on Matisse. In his book *La Morale des Lignes* he
emphasized lines, especially the vertical and the horizontal, and
dreamed of a return to geometry, "mais une géométrie mitigée,
soumise elle-meme à des lois de simplification et d'unification,"
which he thought was important for "le dépouillement de la
réalité dans sa forme la plus abstraite," which in turn was
essential for "la simplification et la modernisation du dessin."[59]
And although he attached subjective sentiments to the vertical
and the horizontal, it is nevertheless these that he thought of as
important in modernizing art. "And is this not already a very

appreciable contribution to artistic evolution and, above all, to the intelligence of contemporary art where the line, presented sometimes without the support of a traditional 'subject,' has to be interpreted and understood by itself and for itself?"[60]

I do not mean to imply that it is uniquely the stimulation of the orientation-selective cells in the brain that results in the aesthetic experience produced by a Malevich or a Barnett Newman, but only that the constituent elements of these works are a powerful stimulus for these cells and, moreover, that a brain deprived of such cells—either because of blindness during the critical period after birth or through pathological reasons—will not be able to appreciate these paintings at all. Given the importance that lines have assumed in much of modern and abstract painting, and given that lines constitute probably the most basic visual stimulus with which to excite a very important category of cell in the cortex, it is interesting to ask whether the relationship between the two is entirely fortuitous.

It is in kinetic art that we find one of the best examples of the art of the receptive field, and its evolution powerfully shows how an art form became better tailored to the physiology of a specific visual area in the brain—area V5, in which visual motion is emphasized.[61] Kinetic art was born out of a dissatisfaction, ostensibly due to social and political reasons, with an art form that seemed to exclude movement or the fourth dimension, as Naum Gabo called it. The demand for its inclusion was strongly featured in the *Futurist Manifesto* of Naum Gabo and Antoine Pevsner and in Marinetti's *Manifesto of Futurism*. For all the shrill demands, especially from the Italian artists, movement was usually represented statically, as in Giacomo Balla's paintings or those of Umberto Buccioni. There are a few early exceptions, such as Gabo's *Kinetic Sculpture,* but they are rare. Marcel Duchamp, influenced by the chronophotography of Jules-Etienne Marais in France, began to produce paintings that suggested movement statically; of these the most famous is perhaps *Nude Descending the Staircase II*. From about 1910 onwards, motion was very much on Duchamp's mind although he did not exploit it explicitly, perhaps because he did not know how to do so or had not yet settled on the best way of doing so. Perhaps Rickey is right in asserting that "Duchamp showed, by deferring his work with movement for years and confining it to optical phenomena, that his concern therein was dadaist and superficial."[62] At any rate, by 1913 he had produced his famous *Bicycle Wheel,* the "Ready-Made," which he called a *Mobile.* Although usually immobile when exhibited in an art gallery, it is commonly thought of as a precursor to kinetic art, even though Duchamp himself did not consider this or machines in general to be artistic objects, referring to them as "non-art."[63] Indeed, the *Bicycle Wheel* was to him only one ready-made among many, which included such interesting objects as urinals—"art without an artist," he called it, a concept that later was to be exploited so well commercially by Andy Warhol, who, it is said, showed the world that anything could be famous for fifteen minutes. The real incorporation of motion in Duchamp's hands came much later, when he produced his *Rotoreliefs* in the 1920s.

Duchamp was not alone in trying to emphasize motion, but the gulf between the idea and its implementation in works of art was not much easier for other artists; it required some degree of technical mastery, of getting at least parts of the work of art into motion, which is perhaps one reason why its actual incorporation into works of art was to take a relatively long time. The surrealists, too, who desired a retreat from all that was rational and predictable, found in motion the unpredictability for which they had yearned. Picabia designed imaginary machines, such as his *Machine tournez vite* and his *Parade amoureuse,* the latter somewhat reminiscent of Duchamp's *La mariée* and, like it, lacking the real motion that it exalted. Until Calder invented his mobiles, the generation of motion depended upon machines, and machines did not seem beautiful or desirable works of art to everyone, not even to the cynical Duchamp.

It was in fact Alexander Calder who best developed the art of the mobile, popularized it, and planted it in the popular mind. In many ways, the mobile was an ingenious invention. It was not dependent upon any profound knowledge of motors and engineering, although Calder's first mobiles were power driven. Mobiles, in other words, were relatively easy to execute. Motion was the dominant element, and to aid the dominance Calder decided to limit himself largely to the use of black and white, the two most contrasting colors, as he called them. Red was to him the color best opposed to these two, but all the secondary colors "confused" the clarity of the mobiles.[64]

One of the specializations in the human visual brain is that for visual motion. This specialization is centered on area V5, where all cells are selectively responsive to motion and the great majority are also selective for the direction of motion, responding vigorously when the stimulus moves in one direction but remaining silent or even being inhibited when it moves in the opposite or "null" direction. These cells are indifferent to the color of the stimulus and usually indifferent to the form as well; indeed, most of them respond best when the stimulus is a spot that is a fraction of the receptive field size. It is interesting to consider here how the mobiles of Calder stimulate the cells of area V5. Viewed from a distance, each element of the mobile is a sort of spot, small or large, depending upon its size. Once it moves in the appropriate direction within the receptive field of a cell in V5, it will lead to a vigorous response from it. In a mobile, of course, the different elements will move in different directions, and each element will stimulate not one but many cells, each cell (or group of cells) being specifically tuned to respond to motion in the respective direction in which the element of the mobile is moving. There are many other interesting features about our perception of mobiles that I have discussed elsewhere,[65] but the important point to emphasize here is that in its development, kinetic art, especially in the hands of Calder, resulted in works that act as perfect stimuli for the cells of V5. Another important feature that perhaps reinforces the view that I present here—that artists try to learn something about the organization of the visual brain, though with techniques unique to them—is found in the general emphasis on movement and in a lack of emphasis on color and form, mirroring so well the physiology of V5.

In giving the above two examples, it is worth emphasizing that there is much about the perception of lines and of motion that we still do not understand physiologically, and it is therefore impossible to relate directly the experience of even one line to what really happens in the brain. If viewed at a sufficiently close distance, even a single vertical line, for example, may fall on the receptive fields of many cells that are specific for the vertical orientation; how the brain combines the responses of these cells to indicate a continuous vertical line is a mystery that neurology has not yet solved, and nor has it solved the question of how it may differentiate one vertical line from other vertical lines that are distinct from it and indeed differentiate the entire tableau from what surrounds it. No less puzzling is the coherence in a work of kinetic art, where the brain can interpret different elements that fall on different receptive fields as forming part of the same work. However, these unsolved neurological problems should not inhibit us from noting that what the physiologists call the building blocks of form—the oriented lines—are the same ones that artists, keen on representing the constant elements of form, have used and that what physiologists consider to be the building blocks in the perception of motion—the cells that respond to motion in a given direction—are the very ones used by an artist such as Calder in his mobiles.

VII.

Jean-Paul Sartre was quite ecstatic about the work of Calder. He wrote, "La sculpture suggère le mouvement, la pienture suggère la profondeur où la lumière. Calder ne suggère rien: il attrape de vrais mouvements vivants et les façonne. Ses mobiles ne signifient rien, ne renvoient à rien qu'à eux mêmes: ils sont, voilà tout; ce sont des absolus."[66] This is not an uninteresting observation, and one can draw at least a superficial parallel between his absolutes and the absolutes of form that were such an obsession of Mondrian and others. The search for these absolutes leads to abstraction. Abstraction has been used to describe many different schools and movements; I use it here in its broadest sense, to signify works in which neither the work itself nor its constituent parts represent any recognizable objects in the visual world (noniconic abstraction). It is obvious that in this context, abstract art differs radically from representational art. The question is whether there is a significant difference in the pattern of brain activity when subjects look at abstract as opposed to representational art?

A hint that there may be substantial differences can be found in recent imaging experiments from this laboratory, which have been inspired in part by the fauvist dream of liberating colors to give them more expressive power. But from what can color be "liberated"? It is not easy to liberate it from form for good physiological reasons. The fauvists therefore settled on a different solution, which was to invest objects with colors that are not usually associated with them, as André Derain's *View of Charing Cross Bridge* and other fauvist paintings testify. Unknown to them, and only uncovered in our imaging experiments, they were exploiting different neurological pathways in the visual brain than the ones used in representational art, where objects are vested in the "correct" colors.

Color is a biological signaling mechanism that exemplifies very well the brain's quest for knowledge under continually changing conditions. It is common knowledge that the basis of color vision is found in light—which itself has no color, being electromagnetic radiation—having many different wavelengths, stretching from red (long wave) at one end to blue (short wave) at the other, and the fact that different surfaces have different efficiencies for reflecting light of different wavelengths. What the brain seemingly does is to compare the efficiency of different surfaces for reflecting light of the same wavebands and thus make itself independent of the actual amount of light of any given waveband reflected from a single surface, since the latter changes continually depending upon the illuminant in which the surface is viewed. If the brain assigned a color to a surface as a function of the wavelength composition of the light reflected from it—characterizing it as green when it reflects more green (middle-wave) light and blue when it reflects more blue (short-wave) light, with the dominant wavelength constituting a sort of code that the brain has to decipher—then the brain would be at the mercy of any and every change in wavelength composition reflected from the surface. Instead the brain has evolved an ingenious mechanism, whose neural implementation remains obscure, to take the ratio of light of a given wavelength reflected from the center and the surround. While the precise amount of light of a given wavelength reflected from a surface changes, the ratio of light of that same waveband reflected from the surface and from surrounding surfaces always remains the same. Color is therefore a construction of the brain, an interpretation that it gives to the reflective efficiency of different surfaces for the different wavelengths of light, which is why James Clerk Maxwell referred to color as "a mental science." But to be able to take ratios, there must be a boundary between one surface and the surrounding surface, and that boundary has a shape—hence the impossibility (except in very rare pathological conditions) of divorcing color, and thus liberating it, from shape. Color therefore follows the logic of the brain's operations. André Malraux was right when he drew attention in *Les Voix du Silence* to Cézanne's remark that "Il y a une logique colorée; le peintre ne doit obéissance qu' à elle, jamais à la logique du cerveau," describing it as "cette phrase maladroite [qui] nous révèle pourquoi, sur l'essentiel de son art, tout peintre de génie est un muet," although I would have preferred it if Malraux had said "devrait être muet" instead.[67]

It is obvious that at the ratio-taking, computational stage there are no "wrong" colors. Making a square red is as good as making it blue. Edwin Land's paradigm in studying color vision consisted of an abstract multicolored scene with no recognizable objects, rather like the paintings of Mondrian. When humans view such a scene the increase in regional activity in their brain occurs in area V4, the color center. But colors are not normally viewed in this way; they are instead properties of surfaces and objects. What happens in the brain when humans view colored objects and scenes depends upon whether the objects are dressed in the right or the wrong colors, but in either case it is different from the activity produced by colors in the abstract, as in a Mondrian. If the objects are dressed in normal colors a more extensive part of the brain, including the frontal lobes, becomes active in addition to V4. But if they are dressed in abnormal colors, as in fauvist paintings, a different set of areas (in addition to V4) becomes active.

These results are replete with neurological interest, but in the present context they allow us to draw two interesting conclusions. The first is that abstract paintings in color do not need to recruit those additional brain areas that are mobilized when we view representational art in color. The second is that the fauvists had unwittingly uncovered certain truths about the organization of the visual brain about which they were and remain ignorant— namely, that their art used pathways that are quite distinct from the ones used by representational art that portrays objects in normal colors.

VIII.

I have tried, using only a few examples, to explain that we have learned enough about the visual brain in the past quarter of a century to begin to study the biological foundations of aesthetics. Aesthetics, like all other human activities, must obey the rules of the brain of whose activity it is a product, and it is my conviction that no theory of aesthetics is likely to be complete, let alone profound, unless it is based on an understanding of the workings of the brain. There is, of course, much that has been left unsaid in this brief essay—about topics such as portrait painting, impressionist art, or op art—but these different tendencies can also be discussed within the overall context of a search for knowledge. There are other questions that are difficult to write about at present: why some artists are drawn to paint in a particular genre, why some of us prefer certain schools to others, the role of the imagination in producing works of art, the relationship between artistic creativity and sexual impulses (since they are both reproductive processes), the emotive power of works of art, the role of culture and historical knowledge in appreciating and interpreting works of art. But I have been exploring a topic here that is new and have concerned myself exclusively with the perceptive aspects. There is much that has yet to be discovered and described.

The approach that I have adopted may seem distasteful to some. Art, they might say, is an aesthetic experience whose basis is opaque and indeed should remain so. It has derived much of its value from the different way in which it arouses, satisfies, and disturbs different individuals, and to profane physiologically the secrets of fantasy in this way implies that what happens in one brain is very similar to what happens in other brains when we view works of art. There is substance to that argument. But we should consider that what happens in different brains when we view works of art *is* very similar, at least at an elementary level, which is one reason why we can communicate about art and through art without the need for the written or spoken word. And no profound understanding of the workings of the brain is likely to compromise our appreciation of art any more than our understanding of how the visual brain functions is likely to compromise the sense of vision. On the contrary, an approach to the biological foundations of aesthetics is likely to enhance the sense of beauty—the biological beauty of the brain.

ACKNOWLEDGMENTS

I gratefully acknowledge the help I received from the staff of the J. Paul Getty Museum in California while I was a visiting museum scholar there at the kind invitation of Mr. John Walsh, Director of the Museum. I am also much indebted to Professor K. Bartels and to Andreas Bartels for their insightful comments, especially concerning the Platonic doctrines.

ENDNOTES

[1] Gerstle Mack, *La Vie de Paul Cézanne*, quoted by Christopher Gray in *Cubist Aesthetic Theories* (Baltimore, Md.: The Johns Hopkins Press, 1953).

[2] Naum Gabo, *Of Divers Arts* (New York: Pantheon Books, 1962).

[3] Paul Flechsig, Gehirnphysiologie und Willenstheorien, Fifth International Psychology Congress (1901), 73–89, trans. Gerhardt von Bonin in *Some Papers on the Cerebral Cortex* (Springfield, Ill.: Thomas, 1960).

[4] M. Vialet, "Considérations sur le centre visuel cortical à propos de deux nouveaux cas d'hémianopsie suivis d'autopsie," *Archives d'Ophtalmologie* 14 (1894): 422–426.

[5] A. Monbrun, "Les affections des voies optiques rétrochiasmatiques et de l'écorce visuelle," in Baillart et al., eds., *Traité d'Ophtalmologie*, vol. 6 (Paris: Masson, 1939).

[6] These are the terms of neurologists, not mine; they were current until the last two decades.

[7] Semir Zeki, *A Vision of the Brain* (Oxford: Blackwell, 1993).

[8] Henri Matisse, *Ecrits et propos sur l'art* (Paris: Hermann, 1972), 365.

[9] Semir M. Zeki, "Functional Specialization in the Visual Cortex of the Rhesus Monkey," *Nature* 274 (1978): 423–428.

[10] Ibid.

[11] Margaret S. Livingstone and David H. Hubel, "Anatomy and Physiology of a Color System in the Primate Visual Cortex," *Journal of Neuroscience* 4 (1984): 309–356. Margaret S. Livingstone and David H. Hubel, "Connec-

tions Between Layer 4B of Area 17 and the Thick Cytochrome Oxidase Stripes of Area 18 in the Squirrel Monkey," *Journal of Neuroscience* 7 (1987): 3371–3377. S. Shipp and Semir Zeki, "Segregation of Pathways Leading from Area V2 to Areas V4 and V5 of Macaque Monkey Visual Cortex," *Nature* 315 (1985): 32–325.

[12]Zeki, *A Vision of the Brain.*

[13]Konstantinos Moutoussis and Semir Zeki, "A Direct Demonstration of Perceptual Asynchrony in Vision," *Proceedings of the Royal Society of London*, Series B 264 (1997): 393–399.

[14]Semir Zeki, "Parallel Processing, Asynchronous Perception and a Distributed System of Consciousness in Vision," *The Neuroscientist* (forthcoming).

[15]Semir Zeki, "A Century of Cerebral Achromatopsia," *Brain* 113 (1990): 1721–1777.

[16]Albert Gleizes and Jean Metzinger, *Cubism* (London: Fisher Unwin, 1913).

[17]Henri Matisse, "Notes d'un peintre," *La Grande Revue* LII (24): 731–745. Reprinted in Jack D. Flam, ed., *Matisse on Art* (Oxford: Phaidon, 1978).

[18]Jacques Rivière, "Present Tendencies in Painting," *Revue d'Europe et d'Amérique* (March 1912): 384–406. Reprinted in Charles Harrison and Paul Wood, eds., *Art in Theory* (Oxford: Blackwell, 1992).

[19]Herbert Read, *The Philosophy of Art* (London: Faber and Faber, 1964).

[20]This does not represent the view of all ancient Greeks; Aristotle, Plato's student, turned away from it.

[21]The example that Plato gives, that of a couch, is derived from the kind of furnishing used in the symposia frequented by Plato and his elite circles. The idea is created by God, the craftsman (δημιουργος) makes a first example of it, and artists subsequently represent different single views of the craftsman's creation.

[22]Arthur Schopenhauer, *The World As Will and Idea*, 3d book, from Albert Hofstader and Richard Kuhns, eds., *Philosophies of Art and Beauty* (Chicago, Ill.: University of Chicago Press, 1964).

[23]John Constable, *Syllabus of a Course of Lectures on the History of Landscape Painting* (London: Royal Institution of Great Britain, 1836).

[24]B. Russel, *History of Western Philosophy* (London: Allen and Unwin, 1946).

[25]G. W. F. Hegel, *Aesthetics*, vol. I, trans. T. M. Knox (Oxford: Clarendon Press, 1975).

[26]Ibid. Emphasis added.

[27]See, for example, Marius von Senden, *Space and Sight* (London: Methuen and Co., 1932).

[28]David H. Hubel and Torsten N. Wiesel, "The Ferrier Lecture—Functional Architecture of Macaque Monkey Visual Cortex," *Proceedings of the Royal Society of London*, Series B 198 (1977): 1–59.

[29]H. Harlow, "Love Created—Love Destroyed—Love Regained," in *Modèles Animaux du Comportement Humain*, no. 198 (Paris: Editions du Centre National de la Recherche Scientifiques, 1972), 13–60.

[30]Semir Zeki and Balthus, *La Quête de l'Essentiel* (Paris: Les Belles Lettres, 1995).

[31]L. C. Perry, "Reminiscences of Claude Monet from 1889–1909," *American Magazine of Art* XVIIII (1927), quoted by John Gage, *Colour and Culture* (London: Thames and Hudson, 1993).

[32]N. K. Logothetis et al., "Shape Representation in the Inferior Temporal Cortex of Monkeys," *Current Biology* 5 (1995): 552–563.

[33]Fritz Novotny, "Das Problem des Menschen Cézannes im Verhaeltnis zu Seiner Kunst," *Zeitschrift fur Aesthetic und Allegemeine Kunstwissenschaft* 26 (1932): 278.

[34]Mack, *La Vie de Paul Cézanne*, quoted in Gray, *Cubist Aesthetic Theories*.

[35]Daniel Henry Kahnweiler, *Juan Gris: Sa Vie, Son Oeuvre, Ses Écrits* (Paris: Gallimard, 1946), 326. Emphasis added.

[36]Ibid.

[37]Ibid.

[38]Rivière, "Present Tendencies in Painting," in Harrison and Wood, eds., *Art in Theory*. Emphasis in original.

[39]Piet Mondrian, *The New Art—The New Life: The Collected Writings of Piet Mondrian* (Boston, Mass.: G. K. Hall, 1986), 338–341.

[40]Kasimir Malevich, *The Non-Objective World*, trans. Howard Dearstyne (Chicago, Ill.: P. Theobald, 1959), 102. Emphasis in original.

[41]Nilsen Laurvik, *Is It Art? Post-Impressionism, Futurism, Cubism* (New York: The International Press, 1913).

[42]Marcel Proust, *Vermeer de Delft* (Paris: La Pléade, 1952), 128.

[43]Paul Claudel, *L'oeil écoute* (Paris: Gallimard, 1946), 240.

[44]Claudel used the English terms since there is no good French equivalent.

[45]Here I disagree with Claudel, who says that the spectator is immediately invited in. This is true of some but not most of Vermeer's work; a notable exception is *Portrait of a Young Girl*.

[46]Semir Zeki, in conversation with Balthus, *Connaisance des Arts*, 477.

[47]Schopenhauer, *The World As Will and Idea*.

[48]Edward Snow, *A Study of Vermeer* (Berkeley, Calif.: University of California Press, 1979), 214.

[49]Gleizes and Metzinger, *Cubism*.

[50]I have used the translation by Symonds; other translations do not use the word brain. The actual word used for brain in the original is *intelleto*. In Latin, *intellectus* means perception or "a perceiving"—see R. J. Clements,

Michelangelo's Theory of Art (New York: New York University Press, 1961), 15—and Symonds has astutely, in my view, rendered it into "brain."

[51]Plotinus, *Ennead V, Eighth Tractate: On the Intellectual Beauty,* republished in Hofstadter and Kuhns, eds., *Philosophies of Art and Beauty.*

[52]Guillaume Apollinnaire, *Les Peintres Cubistes: Méditations Esthétiques* (Paris: Berg International, 1986). Apollinaire does not use the term *Einfühlung.* The notion of *Einfühlung* in art was first elaborated by the German philosopher Robert Vischer in a work entitled *Über das optische Formgefühl.* Wilhelm Worringer developed the notion further and applied it to abstract art in his doctoral thesis at Berne University, published in 1908, entitled *Abstraktion und Einfühlung,* but Worringer sought other, nonneurobiological, explanations for the then developing abstract art. See D. Vallier, *l'Art Abstrait* (Paris: Librarie Générale Française, 1980).

[53]Semir Zeki, "The Woodhull Lecture: Visual Art and the Visual Brain," *Proceedings of the Royal Institute of Great Britain* 68 (1997): 29–63. A more detailed account is given in my forthcoming book, *Behind the Seen.*

[54]Mondrian, *New Art—New Life,* 338–341. Emphasis in original.

[55]Ibid., 75–81.

[56]Ibid.

[57]Kahnwiler, *Juan Gris.*

[58]Gleizes and Metzinger, *Cubism.*

[59]Mécislas Golberg, *La Morale des Lignes,* quoted in Ellen C. Oppler, *Fauvism Reexamined* (New York: Garland Publishing, 1976), 413.

[60]Pierre Aubery, "Mécislas Golberg et l'art moderne," *Gazette des Beaux Arts* 66 (1965): 339–344.

[61]For a more detailed treatment of this subject, see Semir Zeki and M. Lamb, "The Neurology of Kinetic Art," *Brain* 117 (1994): 607–636.

[62]George W. Rickey, "The Morphology of Movement: A Study of Kinetic Art," *Art Journal* 22 (1963): 220–231.

[63]E. Lebovici, "Bouge, moeurs et réssuscite," *Art studio* 22 (1991): 6–21.

[64]Alexander Calder, extract from Julien Alvard and Roger van Gindertael, eds., *Témoignages pour l'art abstrait* (Paris: Editions Art d'aujourd'hui, 1952).

[65]Zeki and Lamb, "The Neurology of Kinetic Art."

[66]Jean-Paul Sartre, "Situations III," trans. Wade Baskin, in *Essays in Aesthetics* (London: Peter Owen Ltd., 1964).

[67]André Malraux, *Les Voix du Silence* (Paris: La Pléiade, 1951), 344.

Richard S. J. Frackowiak

The Functional Architecture of the Brain

INTRODUCTION

W E LIVE IN AN AGE in which people are increasingly accepting the proposition that our emotional, intellectual, and indeed biological lives are determined uniquely by our brains. Nevertheless, as though we were yet in a time when the seat of the soul was thought to be the heart or the pineal gland, we continue to discuss our emotions, intentions, and acts of will in terms that deliberately eschew any reference to the biology of the brain. A popular intellectual pursuit of our time is that loosely termed "neurophilosophy," which attempts to formulate ideas about our humanity by reconciling biological facts about the brain with abstract and historical concepts about behavior. There are a number of reasons why this program has proceeded slowly. A major limitation, imposed by a lack of knowledge, has been the formulation of attributes of human behavior in biological and especially evolutionary terms. A second, related limitation has been our relative ignorance of the physiology and biochemistry of the brain, even of its functional anatomy. This technical impotence has been reversed with the introduction of powerful noninvasive brain monitoring methods. We now have the ability to collect data about how our brains produce all the complex facets of human behavior. It is my belief that, as a result, we will develop a novel, organic view of conscious life—banishing to oblivion endless and essentially

Richard S. J. Frackowiak is Professor of Neurology at the Institute of Neurology, University College London.

specious discussions about whether the brain is capable of investigating itself.

That this enterprise is of paramount importance in modern society is beyond debate. As our populations age and our awareness increases about degenerative diseases that lead to progressive destruction of the personality and humanity of individuals, we confront the fact that we must understand the brain better to have any hope of counteracting the individual and social impact of these diseases. There are those who are frightened of this perspective. The possibility that one day we might be manipulated by malevolent forces through an improved knowledge of brain mechanisms worries scientists and nonscientists alike. But the manipulation of social conscience and individual aspirations is already a fact of life. The impact of images through film and television, sound through recordings and radio, propaganda through the popular press, and indeed ideas through literature are all examples of the modification of human thought and feelings that have their impact on individuals through common brain mechanisms. Why a better knowledge of how our brains are functionally organized would result in an increase of malevolent influences on behavior is not obvious. Indeed, to many the opposite might seem more likely. Such issues, however, need to be addressed when knowledge is ours—and not in a climate of ignorance.

Why do we know so little about the functional architecture of the human brain in comparison with other organs of the body, such as the kidney, liver, and heart? The answer to this legitimate and somewhat puzzling question is almost mundane: The brain—like these other organs apparently homogeneous in composition, at least to the naked eye—is much less readily accessible in life. The average brain weighs approximately 1.4 kilograms and is contained within a bony and relatively impenetrable box, the skull. More problematically, the removal of pieces of tissue during life for examination by microscopy and other anatomical and vital techniques can cause impairments of behavior that make such biopsies ethically unacceptable. The science of brain function has therefore largely depended upon inferences made from observations of abnormal behavior in patients and subsequent correlations with the site and size of damage determined from the brain in postmortem examinations. Though remarkably fruitful over the last 120 years in generating hypotheses about how the brain is functionally organized, there is a weak assumption in this approach. Conclusions depend upon the idea that the human brain is composed of functionally discrete, separable modules, which if removed will leave behind the remaining parts of the brain with their functions relatively unchanged. This assumption is now known to be false in certain circumstances. The damaged brain, just like the normal learning brain, can adapt and reorganize functionally. For example, clinical experience shows that patients with paralysis resulting from brain strokes may show spontaneous recovery of function for many months following an initial lesion. Likewise, regions of the brain that receive sensory signals from one body part may accommodate signals from other body parts under appropriate circumstances.

The new noninvasive techniques for monitoring brain function are providing vast collections of new data about human brain organization—so vast that a significant chapter in the history of our self-understanding is being opened. This chapter began in 1972 with the invention of computerized tomography, or the CT scanner, which offered researchers the ability to reconstruct images of a body in transverse sections. These scanners used x-radiation to generate images of the anatomical structure of the brain with remarkable clarity and without necessitating the physical penetration of the skull. The basic principles of scanning, pioneered by Hounsfield and McCormack (who shared the Nobel prize for their work), were rapidly applied in the mid-1970s to the detection of minute quantities of chemicals rendered visible to scanners by tagging with radioisotopes. Such chemicals could be used to follow vital pathways in the brain using a technique known as positron emission tomography (PET), so called because positron-emitting radioisotopes are used for the tagging. For physical reasons associated with the way such isotopes disintegrate and how the emitted radiation is detected, high reliability can be accorded to PET scan images. The images of brain function generated in this way can be matched with anatomical scans, thus providing a means to examine brain function in life.

There have been enormous technical advances in the performance of scanning machines over the last ten years, and new technologies have joined the arsenal. This essay will review the types of information about brain function that are being obtained. To anticipate somewhat, it can be claimed at present that changes in brain activity associated with pure thought, emotion, and cognitive processes are readily and reproducibly measurable. We are being provided with much new data about how our organ of cognition is functionally organized. In short, our brains are now able to reflect upon themselves using rigorous scientific methods and instrumental forms of measurement.

THE AIMS OF NONINVASIVE HUMAN BRAIN MAPPING

Imaging neuroscience attempts a description of how the brain works at the level of brain systems, that is, functionally segregated populations of nerve cells. The brain looks similar over its entire surface (the cortex), but there are variations in the disposition of nerve cells and of layers of cells in the cortex. These differences have been associated with functional specializations in the cortex, but it has been difficult to identify relevant variations because of a paucity of methods for defining them in detail. Brain areas are activated differently during different behaviors. The functional organization of the brain can therefore be described in terms of the relationships between populations of nerve cells, which themselves may be organized into interacting networks and integrated functional systems. This method of description at the systems level addresses how brain functions arise from the physical structure of the brain. It allows us to ask meaningfully a far broader range of questions than was ever before imaginable: How are primary sensations, such as touch and vision, processed by the human brain? Where are the resultant nervous signals mapped? How are percepts generated, recognized, named, and used to guide subsequent behavior? How do patterns of neural activity in certain brain areas that represent evoked sensations interact with brain areas where patterns of neural activity cause muscles to contract and generate movement? What is the cerebral basis for cognitive functions such as memory, language, and emotion? Are they the result of activity in restricted and highly specialized brain areas, or do they depend on interactions between a number of brain areas with more basic or modular functions?

Early sensory input to the human brain (except for the sense of smell) and late motor output areas are organized as sets of separate maps in the cerebral cortex. That much has been known for many years from neurology and animal studies. The limitations of patient studies have been discussed already. Despite the progress they have made possible, it is unclear how far deductions based on experimentation with nonhuman primate brains can be assumed to be true for the brains of humans. Can rules governing functional brain organization determined in one species be transferred to another? There are obvious changes in the size of the brain between species. Certain functions—for example, spoken language and silent speech—are apparently unique attributes of the human brain; certain areas (such as the frontal lobes) are greatly developed in comparison with even our closest evolutionary relatives in the animal kingdom. This observation implies an addition of cortical areas and of specializations with evolution. A wider understanding of the implementation of a function such as language in the human brain may, through comparative paleontological study, provide a clue as to how this complex means of communication evolved and what its biological antecedents were.

Nerve cells (neurons) are grouped together in the brain and communicate with each other by conveying signals in time-dependent patterns (or codes) with an enormous amount of divergence and convergence of signals between them. This organization can be described at various levels or magnifications, both in space and time. Neurons, neuronal groups, and large-scale functionally homogeneous neuronal populations represent different spatial levels of brain organization. Brain electrical activity can last from milliseconds to seconds. For example, an action potential, the basic unit of electrical signal transmission in the brain, or transmission across the connection between nerve cells (the synapse) last a few milliseconds; the evoked potential, an expression of integrated electrical activity from the cerebral cortex, is measured in hundreds of milliseconds; the readiness potential and the delta wave are evoked and spontaneous brain signals that are recorded with electroencephalography (EEG) and last seconds. These different temporal levels of organization have been brought within the range of measurement by new noninvasive techniques, so that a systematic analysis of both the spatial and temporal functional architecture of the human brain is now possible.

DESCRIBING THE BRAIN'S ANATOMY AND FUNCTION

Using CT scanning, it is possible to image the anatomical arrangement of living human brain tissues by capitalizing on the differential ability of such tissues to attenuate X rays directed through them. CT scanning generates structural images, but the contrast between gray and white matter (containing, respectively, the neurons and the connections between them) is limited by their similar capacity to attenuate X rays. Magnetic resonance imaging (MRI) can generate a broad variety of different structural images of the brain. These images are generated by altering scanning characteristics to sensitize the pictures to differences between gray and white matter. The precision of anatomical information that can be obtained, combined with a lack of dependence on ionizing radiation, makes MRI the present method of choice for anatomical studies. Images can be shown as slices through the brain in any desired orientation. Alternatively, images can be rendered by a computer to provide surface pictures of the cortex of the brain, with its valleys (sulci) and folds (gyri). The resolution normally achievable with comfortable scanning times is 1mm³, which is comparable to the normal 3 to 4 mm thickness of the cerebral cortex.

How then can we obtain detailed information about the function of the human brain in life? Regional human-brain function can be investigated with a variety of techniques, each of which provides unique information. Each technique has limitations and strengths, and many are still under development. The principle of positron emission tomography has already been described: radiolabeled tracers are introduced into the brain via the blood stream, and the resulting regional distribution of radioactivity in the brain is recorded by scanning. This distribution and subsequent detection of the tracer provides information about the biological function in which the tracer participates. In functional mapping (activation) studies the usual variable of interest is the distribution of local blood flow, because blood flow supplies energy to the brain and is a reliable index of local firing in nerve cells; the activity of nerve cells is an energy-dependent process.

The temporal and spatial resolution of functional images obtained with PET are, however, limited. Recording signals for tens of seconds is necessary to generate images that, theoretically, have an optimal spatial resolution of approximately 3 mm^3. In practice, the resolution is often worse—typically a sphere with a diameter of 6 to 10 mm. Even in ideal circumstances, events with short, millisecond-range time constants cannot be demonstrated with PET. A technique that helps to overcome some of these shortcomings is functional MRI (in contrast with the use of MRI to obtain structural information), which records changes in successive images that are related to tissue function. The most useful fMRI method is dependent on local brain oxygen levels (BOLD) and is totally noninvasive. Equivalent brain activity tends to produce smaller changes in image intensities with BOLD fMRI than with PET, but there is greater spatial resolution (roughly 3 mm^3). The relatively low sensitivity to changes in local brain activity of the BOLD fMRI method has led to the elaboration of analysis techniques capitalizing on the fact that images can be acquired very rapidly (50 msec per scan). Changes in a behavioral or physiological state induced during scanning can be conveniently yoked to changes of the image signal by repeated fast imaging. A problem with this strategy is that the BOLD signal has a long (and locally heterogeneous) half-life of several seconds. It is this unfortunate fact, rather than the rapidity with which scans can be recorded, that limits the temporal resolution of this method.

Even so, mapping and analyzing very short-timed brain events and transient correlations of activity in different brain regions is possible with techniques such as magnetoenceph-alography (MEG) and electroencephalography (EEG), both of which record spontaneous electrical brain activity. EEG-based measurements are made from electrodes physically attached to the scalp. The sampling of brain activity by these methods is limited by the size of the head to roughly 120 scalp locations, which results in the poor localization of the number and origin of electrical signals from the brain. One way of increasing the detectability of the evoked electrical activity is provided by event-related potential (ERP) mapping. A cognitive or physiological task of interest is repeated, and recording the evoked electrophysiological activity is time-locked to the stimulus or response in some defined way. The records are averaged to maximize the relevant signal (relative to underlying measurement noise) and then mapped. Recently, improvements in the accuracy with which sources of electrical or magnetic activity are localized have been attempted by the integration of such results with PET or fMRI imaging data. The primary goal of electrical methods is to obtain information about the time course of brain activation, particularly in networks consisting of a number of cooperatively activated brain regions.

THE MAPPING OF SENSORY SIGNALS
ONTO THE HUMAN CEREBRAL CORTEX

Our sensory world depends on stimuli that evoke neural activity, which then maps onto primary sensory areas of the brain. We have some knowledge of these maps, and of the early stages of sensory processing, from experimental results obtained in monkeys and by observations in brain-damaged humans. The visual system has been relatively well studied by some of the modern neuroimaging methods described above and will be used here to illustrate some general principles. A simple imaging experiment to imagine is the measurement of the distribution of brain activity during an eyes-open and an eyes-closed state. The comparison of brain activities in these two states shows the areas of the brain that are activated in association with vision. When early sensory processing is the object of study this comparative approach is relatively free of assumptions; but when more complex cognitive functions are studied, more sophisticated experiments and analyses must be used.

The visual world is mapped from retina to cortex; it is produced by patterns of light hitting the retinas of the eyes. The evoked retinal signals are transmitted to the visual cortex in a point-to-point manner. Signals coming from adjacent patches of the retina (and hence parts of the visual field) are mapped onto adjacent patches of the cortex, a fact that was known from studies of patients with focal lesions of the visual cortex. This retinotopic organization of the primary visual cortex (also known as visual area V1) has been clearly confirmed with scanning in normal humans. Activity recorded in the brain with visual targets in the periphery of vision can be compared to that recorded with a central presentation of the same target. Each quadrant of the visual field is located in the opposite cerebral hemisphere and quadrant of the visual cortex. The point of visual fixation is represented at the pole of the occipital cortex at the back of the brain while peripheral vision is represented in front, at the forwardmost part of the fissure that contains the rest of the primary visual cortex (figures 2 and 3, plates 4 and 5). One can calculate from such data the magnification factor—that is, the length of the cortex that maps a given "length" of the visual field.

FUNCTIONAL SPECIALIZATION IN THE OCCIPITAL CORTEX

The extrastriate occipital cortex (that which lies outside the visual cortex proper) receives an output from visual area V1. This cortex is functionally heterogeneous: different parts are active in conjunction with different visual percepts (for example, form, color, and movement). Visual area V5 is one of these extrastriate areas; activity in it is associated with perceived visual motion. It has been mapped using a visual target in motion and then activity in the brain is compared with that produced when the same target is stationary. Area V5 is found in a circumscribed part of the occipital lobe in front and to the side of area V1 (figure 4, plate 6). When moving objects of different colors and shapes are viewed and attention is drawn to the movement, rather than to the color or shape, activity in area V5 is augmented. The attentional process is implemented by an increase of the response of V5 neurons to the same incoming visual stimulus. How this increased responsiveness is mediated, and whether it depends on the quality of visual signals coming to the brain or on influences altering V5 activity from other brain centers, is crucial to an understanding of attentiveness.

The brains of individuals vary one from another quite markedly, not only in shape, but also in the disposition of the folds and fissures of the parts of the cortex. However, accurate coalignment of functional and anatomical images is a trivial issue with the use of modern computers. It is possible to show precise relationships between structure and function despite considerable anatomical variability in the normal occipital cortex between individuals. Area V5 is always found in relation to two folds of the human occipital cortex. This anatomical site is also characterized by being relatively developed in infants; the nerve connections are well myelinated, or insulated. In summary, there is a remarkable correlation between developmental factors, functional specialization, and anatomical location, despite considerable variability in the absolute spatial location of the anatomical structure in different individuals. Attempts to demonstrate the heavy myelination characteristic of area V5 when examined under the microscope by high-resolution anatomical MRI in life have been partially successful, which is remarkable given that these features are only a millimeter in thickness. The stria of Gennari, a unique structural characteristic of the primary visual cortex (area V1), has also been demonstrated. An informative analysis of structure-function relationships in a normal living brain certainly requires this degree of spatial resolution.

Human color perception, in the sense of seeing a red rose as red whatever the ambient illumination, is associated with the activation of another distinct part of the occipital cortex. This area, known as human area V4, also lies in front of area V1 on the underside of the occipital lobe on both sides of the brain. It has proved possible to correlate the position of area V4 in normal subjects with the damage produced by strokes that lead to a selective inability to perceive color in people who previously had normal color vision. In conclusion, neuroimaging studies of this cardinal sensory system indicate that different visual areas are functionally segregated in the occipital cortex.

We come then to the question of whether the neuroimaging methods now in use and being developed are sensitive enough to detect maps in higher areas of the sensory pathways. The reconstitution of a unique visual percept from activity in functionally segregated, anatomically distinct areas of the brain is a central problem for vision research—and, moreover, for a broader understanding of the biology of perception. The problem of integrating segregated signals and forming a unitary visual percept is more difficult to address at present. One approach has been to use stimuli that generate illusions—for example, form percepts from moving stimuli and motion percepts from nonmoving forms. These stimuli are interesting because among them there are some in which the brain elaborates visual attributes that are not physically present in the stimuli themselves and hence in the real world. They therefore speak to interactions between perceptually specialized areas. Activation of brain areas that include visual motion area V5 by a motion-from-form stimulus shows that unique patterns of visual input do sometimes activate area V5 in the absence of real motion, and that such V5 activation is associated with the perception of nonexistent visual motion. This raises the important idea that the brain *generates* percepts rather than "analyzes" or "interprets" them—and that at times these are elaborated "beyond the information given," resulting in illusions.[1]

Anatomy tells us that there are visual pathways (albeit small compared to the major optic radiation that joins the retina and the visual cortex) that reach functionally specialized visual areas directly, bypassing the primary visual area. Activity in area V5 that is dissociated from the normal preprocessing in area V1 has been shown in a patient whose area V1 was destroyed after a head injury. The patient was sufficiently aware of residual but degraded visual motion in an otherwise completely blind visual field to be able, when tested, to describe the presence and direction of stimuli verbally, without error. This observation leads to the rather interesting conclusion that, in humans, significant visual signals can reach the functionally specialized visual cortex directly. Presumably these signals may also be fed back to area V1 to modulate activity there in response to signals arriving by the classical route. This is an interesting example of a potential control mechanism that involves preparing area V1 for a volley of sensory signals it is about to receive by the major pathway coming to it. Secondly, the perception of visual motion (albeit a degraded form of perception) is, at least in part, a property of the activity in the functionally specialized area alone. Finally, signals from the functionally specialized cortex acting alone can, in abnormal circumstances, be propagated to "inform" function in other areas of the brain—for example, those associated with the language needed to report the visual motion.

BEYOND THE EXTRASTRIATE CORTEX

The awareness of the position of an object and the knowledge of its physical qualities leading to recognition are two visual cognitive functions that depend, at least in part, on a recognition of the object's shape, color, and direction of motion. Imaging studies suggest that the pathways activated in association with these two attributes of objects overlap substantially, but there is also some segregation relevant to each attribute in brain areas forward of the occipital cortex. Activation of posterior parts of the inferior temporal lobes (adjoining the lower occipital cortex) occurs when objects are recognized (for example, to be named). Identification of an object's position in space preferentially activates posterior parts of the parietal lobe (which lies above the temporal lobe and in front of the upper part of the occipital lobe). A third pathway, in which activity is associated with visually guided reaching for objects, has been demonstrated in the parietal lobes between those areas activated by recognition and those by awareness of position. The recognition of further pathways is to be expected because, in general, integration of visual signals with behavior occurs at multiple anatomical levels in the human brain. Each specialized brain area that has connections to another specialized area receives signals back. Each area sends and receives signals to and from multiple other dispersed areas and draws on signals from these areas as the behavioral context demands. Yet signal traffic is not chaotic, a remarkable and often ignored result provided by functional neuroimaging.

Motor associated cortices are functionally specialized and organized into a nested hierarchy of areas comprising a widely distributed system. The brain's representation of simple movements is also organized in maps, so that movements of different parts of the body lie in a reasonably ordered strip along the motor cortex (known as somatotopy). Illustrations of a deformed "homunculus," in which the amount of cortex devoted to sensation or movement is reflected by the size of the body part, are well known. For example, the lips and fingertips are very sensitive and mobile, with correspondingly large cortical representations. Activated areas actually tend to overlap to some extent, but the centers of mass of such activations are clearly separated along each side of the major central fissure of the brain in which the sensory and motor cortices lie.

Somatotopy has also been described in other motor areas that can be found in the front of the brain. Such multiple motor representations have been demonstrated by a selection of appropriate motor tasks during scanning. Brain activations related to movement have been found in at least fourteen parts of the brain, including the primary motor area. A pertinent question is why so many representations of action in the brain can be detected. The most extensive activation of motor-related areas is found when actions are self-selected. Choosing a movement of one's own volition is clearly a complicated process involving cooperation between a multitude of separate, functionally specialized brain areas. Activations of gray-matter structures in the base of the brain (especially two nuclei called putamen and thalamus) are most evident when movements are self-paced or constant, as opposed to when repetitive movements are made. We will next discuss whether the imaging techniques described previously are as informative in studies of more complex functions of the brain as they are in the case of early sensory perception or execution of a motor action. Such complex functions at times appear mysterious because there is no easy way to access them scientifically in animals who lack the ability to communicate them in detail. In short, can the physiological basis of human thought and planning be brought under scientific scrutiny by imaging neuroscience?

Imagining a complex or skillful movement can help improve its performance, a fact well known by musicians and athletes. Brain activity associated with imagined actions can be compared with that at rest or during preparation of a motor act or with its actual execution. The brain areas involved in motor imagery surround those areas in which activity is associated solely with the execution of a motor act. Imagining a complex arm movement is associated with the activation of a number of brain regions in front of and behind the primary motor areas (the premotor and parietal cortices identified in figure 2). Execution of the same complex motor task activates these areas, which are thought to program the movement, and the additional areas that form the executive core" of the motor system and are centered on the primary motor cortex. Such simple experiments illustrate dramatically that it is possible to image brain activity associated with "pure thought"; hence, the idea that thought and introspection can be physiologically studied is, at least in principle, realized.

What brain structures are involved in choosing movements? When brain activity during imagined movements is compared with that at rest, prominent additional activations to those described above are found. Some of these activated areas appear to be specialized for the initiation and selection of movements from a mental repertoire. They are almost exclusively located in the prefrontal cortex near the front of the brain (figure 2). Such areas play supramodal roles, performing functions that are pertinent to memory and other cognitive aspects of action. In particular, these areas are activated when self-generated and stimulus-driven actions are compared. This result is obtained by scanning identical tasks in terms of sensory stimulation and motor response, while giving subjects different instructions during different scans. They are told to do different things with the same stimuli—either responding in a predetermined way or in a self-selected, free manner. In other words, only the nature of the mental operation involved is different during different scans. In this instance it is difficult not to conclude that the activated brain areas, many of them again in the frontal lobes, are directly responsible for the engagement of purely cognitive functions. A description of the physical basis of pure mental activity with even this degree of precision is astonishing and poses challenging questions to philosophers interested in the nature of the "mind."

Indeed, neuroimaging data can be used to investigate functional interactions between brain regions. When scans collected during self-generated and stimulus-driven action are compared, there are frontal activations as described above, whatever the precise detail or modality of the sensory stimulus or action involved. However, modality-specific changes in brain activity are also found. They are usually deactivations when self-generated task scans are compared to stimulus-driven ones. This pattern of response is opposite to that found when stimulus-driven behavior is examined; in that case, modality-specific brain areas show enhanced activity. This constitutes evidence of correlated activity linking task-dependent, modality-specific areas usually found in the back half of the brain and amodal (hence higher) areas located in the frontal cortex. Correlated activity implies a functional interaction between neuronal populations, and the images provide data that can be used to test and inform population-level models of brain function based on the observed functional connectivity between brain regions. More generally, the frontal cortex, which differs hugely in volume between humans and even our closet primate relatives, is greatly implicated by neuroimaging in processes involving planning, choice, volition, memory, and similar cognitive functions.

In general, multiple scans recorded over time generate the image data needed to describe the spatial distribution of brain activity correlated in time. The volumes of data can be processed using advanced statistical mapping techniques to generate images that reflect functional or effective connections between brain areas. Further, it is possible to dissociate independent systems that interact at common sites in the brain. *Functional connectivity* is the correlation of activity in different brain regions evoked by a stimulus or behavior. Such correlations may be found because of direct or indirect interactions between brain areas or by virtue of common activation from lower (subcortical) structures. *Effective connectivity* is the influence one brain region exerts on another; it can be assessed quantitatively. The ability to measure effective connectivity presents opportunities to measure modifications in the strength of connections between brain areas by allowing us to observe how such interconnections are affected by behavioral manipulations or drugs.

Can cognitive processes be added and subtracted as independent unitary processes? A direct comparison of brain scans in different states is not always appropriate for the study of cognition; such a strategy rests on the assumption that interactions between brain areas and activity in them can be regarded as additive (or subtractable) one with the other. Simple sensory stimulation is associated with correlated activity in primary sensory areas of the brain. Correlations between the activity and the strength of the stimulation are often simple and linear. However, when sensory signals are distributed among brain areas for the purposes of guiding or modifying behavior, the evoked activations are rarely simply correlated with the amount of sensory input. There are two principal reasons for this. Sensory signals may drive activity of a region (bottom-up activation). On the other hand, the response to a volley of sensory signals in a brain region may be altered by top-down influences from yet other areas. These areas may be mediating activations that set a behavioral context, mediate the degree of attention, reflect how well a behavior has been learned, and so on. Such context-modifying signals will interact with the sensory volley in a nonlinear manner and will, in this way, result in interactions between experimental factors that themselves determine the distribution of activity. Techniques are available that take into account such complex, higher-order interactions within and between activated brain regions. They search for nonlinear responses and the context-sensitive modulation of activity in brain regions.

Certain brain processes are activated by tasks despite the absence of explicit instructions designed to engage them. When subjects are asked to recognize particular orthographic features of words rather than to read them, a network of brain areas comprising a large number of anterior and posterior brain regions is activated. Many of these regions are known, from other studies, to activate with a variety of language-related functions, including the appreciation of meaning. There is thus obligatory activation, in the absence of explicit instruction or conscious effort, by word-like visual stimuli of areas in which activity is associated with the appreciation of meaning. It may not always be possible therefore to make assumptions about the constituent cognitive processes engaged by a task.

THE PROBLEM OF SELF-REPORTING

Many functional neuroimaging results depend on the cooperation or the self-reporting by subjects during or after scanning.

This fact is sometimes used to suggest that the data obtained are "soft." Introspection and the reporting of observations or actions is an intimate part of everyday existence, even in the sphere of science. A statement that an object is a meter long is an example of self-reporting by the measurer. We do not doubt the report, because it is consistent, can be repeated by others, and is demonstrably true in practice (for example, reliable working machinery can be made based on the correctness of the measurement). Introspection has been used to investigate human brain function by neurologists and neuropsychologists with the clinico-pathological lesion method for at least a century. The repeated demonstration of consistent patterns of local brain activation during defined mental activity now provides objective measurements that are difficult to refute. Such measurements bring the investigation of thought, consciousness, emotion, and similar brain functions into the realm of "hard" scientific inquiry.

Cognitive tasks dependent on self-reporting are associated with responses that can be recorded or physiological changes that can be measured. Scanning that depends on recording task performance and correlating such results with scan data can be used to identify brain regions in which activation is coupled in some way to a task. The difficulty of the tasks involved may be varied. Such correlated observations remove the need for control scans and potentially circumvent the assumption of pure subtraction. For example, in the visual system, activity in the visual cortex increases with the increasing frequency of a flickering light, reaches an apex, and then falls at fast frequencies when it becomes difficult or impossible to discriminate individual flashes perceptually. Scanning while listening to pure tones of different frequencies results in brain signals that line up along the primary auditory cortex in a "tonotopic" map. One tone played at different volumes can be used to identify its representation in a tonotopic map by correlating brain activity with volume.

Patterns of brain activity are transformed as the signals they process are transmitted to new regions of the brain. The primary auditory cortex is the brain area to which nerves coming from the ear that are associated with hearing project. In the primary auditory cortex, activity increases in proportion to the number of words spoken per unit of time. The posterior temporal cortex, to which the auditory cortex projects and which lies just behind it, shows a different response with the same stimulation; activation is apparent as soon as words are heard, but no further detectable change in local brain activity occurs across a range of word frequencies. The conversion of a rate-dependent response in the primary auditory cortex to one in which activity is not rate-dependent suggests a mechanism for integrating frequency-determined neural activity into a form that signals words having a phonological and semantic identity (sound and meaning).

Similar findings have been observed in the motor system. The activity in the primary motor cortex and associated regions of the "executive" motor system increases exponentially with the rate of repetitive movement or the amount of constant force exerted. There are no such exponential changes in other motor-related areas associated with the initiation or sequencing of movements.

THE LOCALIZATION OF MEMORY

Thus far we have discussed functional neuroimaging in brain systems related to input (sensory) and output (motor) systems. Our discussion has progressed beyond these into the cognitive domain with examples of increasing complexity and an analysis of task-dependent issues relevant to informative experimental design. The final part of this essay will deal with the functional neuroanatomy of human memory, a topic characterized by a large and increasing variety of identified memory processes and the complex interactions between them. This fact makes particular demands on an appropriate choice of both imaging methods and task definition for the attribution of function to structure.

How do we keep snippets of information in mind for short periods (working memory)? Many cognitive processes are composed of a number of subprocesses, some of which can be inferred from the fact that one can find patients who have deficits in one and not another component, and others who have the reverse pattern—an observation known in clinical neuropsychology as "double dissociation." Psychological data based on reaction times in specially designed series of interrelated tasks in normal subjects provide additional information. An influential model of verbal working memory developed by Alan Baddeley and his coworkers incorporates at least two subprocesses—a rehearsal system and a phonological store. The former refreshes the contents of the latter, which acts as a limited buffer (of three to four words) with a half-life of approximately two seconds. A common experience that illustrates what is meant by working memory is remembering a new telephone number before finding pencil and paper to write it down. If interrupted in the process of repeating the number to oneself, the information is lost.

The validity of this conceptual framework has been investigated with neuroimaging. The subprocesses of working memory were identified separately, and activity due to confounding memory processes was accounted for. This was done by controlling for potential confounds by the use of experimental designs in which independent factors are separately varied in different scans. For example, the ability to retain a series of letters in verbal working memory can be controlled by a task in which shapes are remembered that have no phonological connotations during identical conditions of visual presentation. The difference between these scans represents the dimension of verbal memory load. According to the model, an ability to make rhyming judgments is a function that primarily engages rehearsal. A scan during rhyming can be controlled by scanning during judgments of shape identity where (again) no phonological processing—an absolute requirement for making rhyming judgments—is involved. The difference between these scans represents the dimension of phonological load. The interaction between mnestic and phonological dimensions eliminates contributions to the activation pattern from known and unknown interfering processes, while identifying brain areas associated with the two subprocesses and their interactions alone. Such studies have shown a critical role for the inferior frontal lobe at the front of the brain on the left in the rehearsal function, and for the inferior parietal lobe (again on the left) in the functions of the phonological store.

How then do we remember events in our personal lives? Longer-term autobiographical memories available to consciousness (episodic memories) present a particularly interesting object of study. The process of remembering lists of words, for example, can be contaminated by mechanisms that permit recall with above-average success in the absence of explicit learning. Use has been made of the known interference caused by difficult distractor tasks with the acquisition of explicit memories. Scanning can be carried out during the performance of a paired associated-word learning task or a controlled, repetitive passive-listening task with and without concurrent distraction. The passive-listening task controls for auditory and other known and unknown components of the memory task. The effect of distraction on learning eliminates efficient episodic encoding (learning), thus providing a control activation map representing areas associated with priming and other irrelevant processes. (Priming is the well-recognized facilitation of recognition caused by prior exposure to a stimulus that a subject is not deliberately trying to remember.) The difference between this map and that comparing episodic learning with control in the nondistracting state indicates areas specifically associated with the episodic learning process. From such a task design it is possible to show that episodic learning is selectively associated with the activation of a localized posterior midline region of the cortex (retrosplenial) and a region of the left-sided (dorsolateral) frontal cortex that has already been implicated in other cognitive tasks discussed above (figure 5, plates 7 and 8). It is important to remember that, though these areas have a particularly critical importance to normal episodic memory acquisition, memory function remains a property of the whole network.

Anatomical correlates of the recall of previously learned episodic memories have also been discovered. Scanning is performed while subjects view the first of a previously presented pair of words (a category such as countries) and attempt to recall the second of the pair (a specific example, say, "England"). In this task a strategy involving semantic knowledge about the category might contaminate episodic recall, especially if subjects resort to guessing the example if it does not immediately come to mind from the learning period. Scanning can be carried out during a similar task in which a novel (rather than previously learned) series of category words are presented. This task depends entirely on previous knowledge and can be used as a control in experimental episodic recall tasks. A repetition task is used to control out common aspects of listening to words and other less-identifiable subprocesses. A comparison of scans performed during a repetition-control task with scans performed during experimental episodic tasks gives a map of areas involved in both episodic recall and recall from an individual's knowledge base (semantic memory). On the other hand, a comparison of repetition scans with semantic recall scans gives a map of areas specifically associated with semantic memory. The difference between the two results will isolate areas associated with episodic recall alone. The result indicates a prominent role for the right frontal cortex and another, separate posterior cortical area (the precuneus) in the recall of episodic memories (figure 5). This right/left frontal distinction between cerebral activations evoked by remembering and laying down episodic memories is a clear functional specialization in a very high-order cognitive system, and it has been confirmed in many studies using a wide variety of different materials for remembering.

Extending these analytic tools in other ways shows us that learning new skills involves large-scale, time-dependent changes in patterns of neuronal activation. Behaviorally, repetitive task performance results in habituation and adaptation effects. Learning to play a new piece on the piano or to play tennis initially demands great effort—but progressively becomes more automatic, or skillful. The physiological correlates of such general mechanisms have been shown with neuroimaging. For example, in a verbal task requiring responses to novel categories by giving a specific example, there is activation in a distributed network that includes the frontal cortex and a number of language-related areas in the temporal lobe. If such a task is repeated until the responses are overlearned, the pattern of activation is attenuated and resembles that obtained with the simple repetition of words. This result indicates that response selection has become automatic, that is, subjects no longer have to think about what they are doing. Introducing a new category target brings about a return to the original pattern of activation.

Motor-skill learning can be measured during the repetitive performance of a manual dexterity task. Following a rapidly rotating target with a hand-held stylus is initially difficult; accuracy at different speeds and with practice can be recorded. Improved performance becomes observed as accuracy and time-on-target increase. Such improvement correlates with increased activity in primary and other motor cortices. Activation of the brain during the performance of such tasks can be modified for two reasons. First, improved performance will result in greater motor activity during a scan that will be due simply to improved performance. Second, modifications of activity may occur due to the acquisition of a new motor skill (or the memory of a motor action). A proper interpretation of the imaging result therefore depends on a realization that improved performance and increased skill are separate, but confoundable, attributes of a motor act. When a motor task is performed repeatedly to the same performance criterion, thus eliminating any performance confound, then progressive attenuation of activity occurs in both the premotor cortex and the cerebellum. (The cerebellum lies underneath the occipital lobe; it is in the skull but is separate from the cerebral cortex above.)

Scanning during the learning of a novel sequence of key presses (such as in learning a piano piece) with error feedback via an auditory signal results in the greater activation of parts of the right frontal cortex when a performance is naive than when the sequence has been overlearned. Conversely, activity in other midline parts of the frontal lobe is greater in the automated state than in the naive state. Visual and language-associated cortices show considerably less activity during the naive state than when a key-press sequence is overlearned. When a task is novel and requires considerable attentional resources, there appear to be mechanisms for large-scale deactivations of whole systems that are not required for the task. When a task becomes overlearned, attentional needs decline and activity in extraneous areas normalizes relative to activity in the remainder of the brain.

Paying attention to stimuli is a general psychological mechanism that generates different patterns of activity depending on which attentional process is engaged. Selective attention, as noted already, results in the augmentation of activity in modality-specific areas specialized for the elaboration of a function or percept to which attention is directed. An example from studies in the visual system has already been given. Recent advances allow the measurement of the influence that activity in one brain region has on other regions. We have shown that in a visual categorization task using objects presented at various rates, areas can be identified in which activity is dependent on the amount of attention required. Difficult (as compared to easy) categorization tasks activate frontal regions. Selective attention to object attributes at different levels of difficulty results in two types of modulation of activation in the inferior temporal cortex. The relationship between difficulty and activity changes so that there is a greater activation for equivalent rates when attention is engaged. The source for this modification of the stimulus-response relationship appears to be in the frontal cortex. In addition, there is a change in the unstimulated activity of the same inferior temporal cortex when the task provides an expectation of increased attentional requirements. These two mechanisms can be considered analogous to a change in the gain characteristic of the object recognition system (analogous to a change in the volume control in a radio set) and a change in its offset with expectation (in this instance, analogous to a change in the wavelength band or channel). This example serves to show that functional neuroimaging provides information about brain mechanisms as well as brain localization.

SYNTHESIS AND CONCLUSIONS

The field of functional neuroimaging is in a state of rapid technical development. The data derived from different scanning methods are often complementary, and there is much evidence to suggest that this state of affairs will continue, thereby providing a wide range of improved tools for the exploration of the functional architecture of the human brain. Neuroscientific progress will be made by judicious use of one or more methods to answer appropriate questions. We can conclude, on the basis of results to date, that it is now possible to embark on research into the functional architecture of the living human brain that goes beyond descriptive "neophrenology" (the attribution of functions to sites in the brain). Nevertheless, it remains a fact that our description of the anatomical arrangement of the functioning human brain is incomplete, and therefore even "neophrenology" remains an important area of continuing study. Uniquely human functions must be assigned to networks of brain areas; cognitive processes require definition in physiological and anatomical terms.

New techniques, such as the measurement of functional connections between brain areas, may have practical significance for the treatment and modification of mental diseases and cognitive function. For example, the symptoms of schizophrenia are modified by drugs working on the dopamine system. Patients show a number of cognitive abnormalities that can be related to disorganized function in well-defined brain systems. The restitution of normal functional relationships in such a system by a dopamine active agent has been demonstrated, and the locus of interaction between dopamine and cognition-related brain activity has been identified. Such basic neurobiological information will generate ideas relevant to drug design and assessment.

Above all, these new methods are leading to better human self-understanding through an appreciation of the unique nature of how our mental activity is implemented in the organic matter that we carry around in our skulls, that enormously complex organ that defines our personalities, hopes, wishes, actions, and ambitions.

ACKNOWLEDGMENTS

The author wishes to acknowledge funding and support from the Wellcome Trust of the United Kingdom.

ENDNOTE

[1]Semir Zeki, J. D. G. Watson, and Richard S. J. Frackowiak, "Going Beyond the Information Given: The Relation of Illusory Visual Motion to Brain Activity," *Proceedings of the Royal Society of London,* Series B, 252 (1993): 215–222.

SELECTED READING

Baddeley, Alan. *Human Memory: Theory and Practice.* London: Erlbaum, 1990.

Binder, J. R., et al. "Functional Magnetic Resonance Imaging of the Human Auditory Cortex." *Annals of Neurology* 35 (1994): 662–672.

Boecker, H., et al. "Functional Co-operativity of Human Cortical Motor Areas during Self-paced Simple Finger Movements: A High-resolution MRI Study." *Brain* 117 (1994): 1231–1240.

Colebatch, J. G., et al. "Regional Cerebral Blood Flow during Voluntary Arm and Hand Movements in Human Subjects." *Journal of Neurophysiology* 65 (1991): 1392–1401.

Courtney, S. M., et al. "Object and Spatial Visual Working Memory Activate Separate Neural Systems in the Human Cortex." *Cerebral Cortex* 6 (1996): 39–49.

Detre, J. A., et al. "Perfusion Imaging." *Magnetic Resonance Medicine* 23 (1992): 37–45.

Frackowiak, Richard S. J. "Functional Mapping of Verbal Memory and Language." *Trends in Neuroscience* 17 (1994): 109–115.

Frackowiak, Richard S. J., et al., eds. *Human Brain Function*. San Diego, Calif.: Academic Press, 1997.

Frith, Christopher D., et al. "Willed Action and the Prefrontal Cortex in Man." *Proceedings of the Royal Society of London,* Series B (1991): 244.

Grafton, S. T., et al. "Somatotopic Mapping of the Primary Motor Cortex in Humans: Activation Studies with Cerebral Blood Flow and Positron Emission Tomography." *Journal of Neurophysiology* 66 (1991): 735–743.

Grafton, S. T., et al. "Human Functional Anatomy of Visually Guided Finger Movements." *Brain* 115 (1992): 565–587.

Hari, R., and O. V. Lounasmaa. "Recording and Interpretation of Cerebral Magnetic Fields." *Science* 244 (1989): 432–436.

Hari, R. "Human Cortical Functions Revealed by Magnetoencephalography." *Progressive Brain Research* 100 (1994): 163–168.

Hillyard, S. A. "Electrical and Magnetic Brain Recordings: Contributions to Cognitive Neuroscience." *Current Opinions of Neurobiology* 3 (1993): 217–224.

Makela, J. P., et al. "Functional Differences Between Auditory Cortices of the Two Hemispheres Revealed by Whole-head Neuromagnetic Recordings." *Human Brain Mapping* 1 (1993): 48–56.

Posner, Michael I., and Marcus E. Raichle. *Images of the Mind*. New York: Scientific American Library, 1994.

Raichle, Marcus E. "Circulatory and Metabolic Correlates of Brain Function in Normal Humans." In Fred Plum, ed. *Handbook of Physiology: The Nervous System,* vol. V. New York: American Physiological Society, Oxford University Press, 1987.

Roland, Per E. *Brain Activation*. New York: Wiley-Liss, 1993.

Rugg, M. D. "Event Related Potentials in Clinical Neuropsychology." In J. R. Crawford, D. M. Parker, and W. W. McKinley, eds. *A Handbook of Neuropsychological Assessment*. London: Erlbaum, 1992.

Seitz, R. J., and P. E. Roland. "Learning of Sequential Finger Movements in Man: A Combined Kinematic and Positron Emission Tomography (PET) Study." *European Journal of Neuroscience* 4 (1992): 154–165.

Stehling, M. K., R. Turner, and P. Mansfield. "Echo-planar Imaging: Magnetic Resonance Imaging in a Fraction of a Second." *Science* 254 (1991): 43–50.

Stephan, K. M., et al. "Functional Anatomy of the Mental Representation of Upper Extremity Movements in Healthy Subjects." *Journal of Neurophysiology* 73 (1995): 373–386.

Tootell, R. B., et al. "Functional Analysis of Human MT and Related Visual Cortical Areas Using Magnetic Resonance Imaging." *Journal of Neuroscience* 15 (1995): 3215–3230.

Tulving, Endel. *Elements of Episodic Memory.* New York: Oxford University Press, 1983.

Wilding, E. L., M. C. Doyle, and M. D. Rugg. "Recognition Memory with and without Retrieval of Context: An Event-related Potential Study." *Neuropsychologia* 33 (1995): 743–767.

Zeki, Semir M. *A Vision of the Brain.* London: Blackwell Scientific Publications, 1993.

Zeki, Semir, J. D. Watson, and Richard S. Frackowiak. "Going Beyond the Information Given: The Relation of Illusory Visual Motion to Brain Activity." *Proceedings of the Royal Society of London,* Series B, Biological Science, 252 (1993): 215–222.

Zeki, Semir, et al. "A Direct Demonstration of Functional Specialization in the Human Visual Cortex." *Journal of Neuroscience* 11 (1991): 641–649.

Mark F. Bear and Leon N Cooper

From Molecules to Mental States

I
T IS REMARKABLE HOW LITTLE we need to know about ourselves in order to survive and reproduce. Humans are probably the only animals that have made conjectures about
what goes on inside the body. And even then, it has taken a long
time to arrive at a clue. It was less than four hundred years ago
that Harvey taught us that the heart was a pump. The great
Aristotle had conjectured that the heart was the seat of intellect, reserving for the brain the function of a cooling system,
which (in addition to some other functions of more than minor
interest) it is. In dealing with questions about our own nature
we have to survive ideology—endless discussions about organic
versus inorganic, vital versus nonvital, living versus nonliving.
These hotly disputed questions quietly fade away when the cool
light of patient investigation and hard thought finally provide
illumination. Today we are engaged in the quest of understanding our brain. When we have finally worked out the details, this
remarkable organ, for all of its lofty pretensions—seat of intellect, home of the soul—will very likely join other remarkable
pieces of biological machinery. Remarkable, certainly, but not
mysterious or possessed of any supernatural qualities.

How then do we go about trying to understand a system as
complex as the brain? Obviously we cannot just make observa-

*Mark F. Bear is professor in the Department of Neuroscience, Associate Investiga-
tor at Howard Hughes Medical Institute, and Associate Director of the Institute
for Brain and Neural Systems at Brown University.*

*Leon N Cooper is Thomas J. Watson, Sr., Professor of Science, professor in the
Departments of Neuroscience and Physics, and Director of the Institute for Brain
and Neural Systems at Brown University.*

tions. The number of possible observations is substantially larger than the available number of scientist-hours, even projecting several centuries into the future; the result might be a listing of facts that would be of little use. A theory or a point of view is essential. These shape the direction of the analysis as well as guide us toward what we believe are the relevant observations. Theory provides a framework within which questions become relevant.

The usefulness of a theoretical framework lies in its concreteness and in the precision with which questions can be formulated. The more precise the questions, the easier it is to compare theoretical consequences with experience. An approach that has been very successful is to find the minimum number of assumptions that imply as logical consequences the qualitative features of the system that we are trying to describe. If we pose the question this way, it means that we agree to simplify. As Albert Einstein once said, "Make things as simple as possible, but no simpler." Of course, there are risks, and it is here that science becomes as much art as logic. One risk is that we simplify too much and in the wrong way, that we leave out something essential. This is the intellectual risk. Another risk is political—that we may choose to ignore in the first approximation some facet that an important individual has spent a lifetime elucidating. (Of course, temporarily setting this facet aside does not reduce its importance. Once we have achieved the initial [zero-order] scaffolding, this facet may provide an essential variation.)

The task, then, is first to limit the domain of our investigation, to introduce a set of assumptions concrete enough to give consequences that can be compared with observation. We must be able to see our way from assumptions to conclusions. The next step is experimental: to assess the validity of the underlying assumptions, if possible, and test their predicted consequences.

This procedure has been extraordinarily successful in physics. There we have mathematical structures, supported by experimental observations, that link the origin of the universe at ferocious temperatures to the behavior of electrons on a laboratory table at the lowest temperatures attainable. However, a legitimate question is whether such procedures can be duplicated for biological systems. Although there are examples of extremely valuable biophysical theories (e.g., the mathematical description of the action potential by Hodgkin and Huxley), the question is often raised whether complex neural systems, involving vast numbers of mutually interacting elements, can be illuminated by theoretical structures. We hope to show in this essay that not only is such illumination possible, but to some extent it has already been realized.

WHAT IS A GOOD THEORY?

What is necessary to make connections between assumptions and conclusions? If one or two steps are all that is required, little mathematics is needed. However, if what is required is a long chain of reasoning with a quantitative dependence on parameters, mathematics, while possibly not required, helps. (It is important to distinguish such mathematical structures from computer simulations, where the preciseness of assumptions is often lost in mountains of printouts.)

A "correct" theory is not necessarily a good theory. It is presumably correct to say that the brain, with all of its complexity, is one consequence of the Schrödinger equation (the basic equation of quantum physics) applied to some very large number of electrons and nuclei. In analyzing a system as complicated as the brain, we must avoid the trap of trying to include everything too soon. Theories involving vast numbers of neurons in all their complexity can lead to systems of equations that defy analysis. Their fault is not that what they contain is incorrect, but that they contain too much.

The theory that predicts everything predicts nothing. Thus a theory seemingly can be characterized by what it does not predict (and thus can be falsified). But this can be illusory. Additional ad-hoc hypotheses can almost always be introduced to save the original idea. (Think of the use of epicycles to modify circular orbits in the earth-centered universe). But ad hoc hypotheses (even though occasionally correct) become increasingly unattractive. Thus we add esthetic criteria: that the theory be beautiful, natural, and elegant with few assumptions, possess a rich structure, and have a detailed agreement with experience.

Most great theories are at best partially correct. The famous Bohr atom was grounded on inconsistent assumptions and was never successfully extended beyond hydrogen. It did, however, give us a stunning, if approximate, account of the spectrum of light emitted and absorbed by hydrogen atoms. One possible position (taken by some) was that because of its deficiencies, it was not worth pursuing. Another (in retrospect, very fruitful) position was that such remarkable agreement between the theoretical structure and what was seen could not be accidental. Thus, the useful question was: What are the consistent new assumptions that yield in the relevant domain, as one consequence, the structure of the Bohr atom?

A theory is not a legal document. In spite of occasional suggestions to the contrary, no scientist is in communication with the Almighty; there are no tablets delivered from Sinai. Theoretical analysis is an ongoing attempt to create a structure—changing it when necessary—that finally arrives at consequences consistent with our experience.

The most important characteristics of a good theory are precision and concreteness. It should be well defined and precise; to paraphrase Galileo, what is said should depend on what was said before. This is not to say that a theory cannot be amended or modified. Indeed, one characteristic of a good theory is that one can move about in the structure—tinker, change assumptions—and know what the new consequences will be.

But in many ways asking what makes a good theory is like asking for a general prescription for a good painting or a good piece of music. We can list many rules (to which there are always exceptions), but in the end you know it when you see or hear it. In what follows we give an account of our attempt to construct and experimentally verify a theory describing an important aspect of brain function: how the behavior of individual neurons changes depending on experience in such a way that memories are stored.

A PROBLEM AND ITS THEORETICAL SOLUTION

Consider the mystery of visual-recognition memory. Humans possess an extraordinary ability to recognize the familiar. The person you met for the first time this morning will be recognized this afternoon, tomorrow, and very likely a year from now. This ability is essential for us to conduct our daily affairs. Indeed, the loss of this type of memory is an early and frightening sign of many dehumanizing neurological diseases, such as Alzheimer's. How are such memories formed and maintained in the brain?

The answer, we believe at present, begins with the neuron—a specialized cell found in great abundance in our head and which is the basic cellular entity responsible for the special properties of the brain. Neurons possess axons, which serve as the wires that conduct electrical impulses from one point in the brain to another. When the impulse finally reaches its destination, the information is transferred to other neurons at specialized sites of interneuronal contact called synapses. How much the presynaptic impulse affects the electrical activity of the postsynaptic cell depends on what is often called the "gain," "weight," or "strength" of the synapse.

One of the early insights in this field was that by appropriately assigning synaptic strengths within a network of neurons, one could construct "memories" with properties strikingly reminiscent of our own. Consider as an example the simple model in figure 6 (plate 9). Three neurons (labeled 1, 2, and 3) receive synaptic inputs that convey information about three stimuli (labeled A, B, and C). Initially, before learning, each neuron responds similarly to each stimulus; there is no pattern of cellular output that uniquely represents each stimulus. After learning, however, the synapses have modified so that different stimuli yield different responses. Note that although each stimulus evokes a maximal response from a different neuron, the neural representation of a stimulus is distributed over all three cells. Stimulus A, for example, evokes a large response in cell 1, a moderate response in cell 2, and a weak response in cell 3. The representation of stimulus A is this unique combination of responses across the cells in the network. Thus, the memory in such a network is said to be "distributed."

Such distributed memories are resistant to the loss of individual neurons. For example, the loss of cell 1 would still leave an activity ratio in cells 2 and 3 that is unique to stimulus A. So-called graceful degradation of memory is also a feature of human aging. Rarely is there a catastrophic loss of particular memories as particular neurons die; rather, all memories become increasingly fuzzy. An analysis of neural networks suggests that this fuzziness results when enough neurons have died so that the signal-to-noise ratio decreases and the representations begin to blend together.

Models of distributed memory storage, such as that in figure 6, suggest that a cellular correlate of memory is experience-dependent changes in neuronal stimulus selectivity. By changes in "stimulus selectivity" we mean that neurons come to respond most vigorously to a subset of their synaptic inputs. Indeed, neurophysiological studies of neurons in the brain have revealed precisely this type of change as animals learn to recognize and discriminate stimuli. One example comes from the work of Edmund Rolls and his colleagues at Oxford. They measured the responses of neurons in the inferotemporal cortex as monkeys learned to recognize new faces. As learning occurred, the responses to the new faces decreased in some neurons and increased in others, reflecting shifts in selectivity among this population of neurons. This learning-related change in "face selectivity" in inferotemporal cortex is of particular interest: in humans, lesions of this region cause a fascinating disorder called prospagnosia, a severe disturbance in the ability to recognize familiar faces. These and many other observations have focused attention on the question of how synapses "learn" under the influence of experience such that neuronal stimulus selectivity is altered.

An early proposal usually ascribed to the Canadian psychologist Donald Hebb has played a major role in an entire class of what are called "unsupervised learning rules" (those that do not require the intervention of an external "teacher"). Hebb suggested that rudimentary associative memories could be formed if synaptic strengths modified based on information available locally at the synapse, namely, the timing and amount of pre- and postsynaptic activity. Serious theoretical analysis followed. In the 1970s, a number of "learning rules" were proposed to account for the experience-dependent development of stimulus selectivity, each making different assumptions about how synapses would modify during various combinations of presynaptic and postsynaptic activity.

One such synaptic learning rule, from the work of Leon Cooper and his colleagues, is illustrated in figure 7 (plate 9).[1] They considered a single neuron receiving an array of excitatory synapses carrying information about the sensory environment. In order to account for the development and plasticity of neuronal stimulus selectivity, they proposed that active synapses are potentiated when the total postsynaptic response exceeds a critical value, called the "modification threshold" (θ_m), and that active synapses are depressed when the total postsynaptic response is greater than zero but less than θ_m. It is readily apparent how this rule leads to stimulus selectivity. Consider the situation in figure 2, where postsynaptic responses to stimuli A, B, and C are initially clustered around the value of θ_m. Because the response to C is greater than θ_m, the synapses that are active during the presentation of C potentiate. Likewise, because the responses to A and B are less than θ_m, the synapses that are active during the presentation of A and B depress, and the neuron becomes selectively responsive only to stimulus C.

While this synaptic learning rule does yield stimulus selectivity under the appropriate conditions, a fixed θ_m can have undesirable consequences. For example, if patterns A, B, and C all initially yielded responses below θ_m, then all synapses would depress and the cell would cease responding to any stimulus. On the other hand, if patterns A, B, and C all initially yielded responses greater than θ_m, then all synapses would potentiate to their saturation limit and the cell would lack selectivity. A solution to this problem was introduced by Bienenstock, Cooper, and Munro in what is known as the BCM algorithm.[2] They showed that if the value of θ_m was allowed to vary as a nonlinear function of the average integrated postsynaptic activity, then the cell would evolve to a stable, selective state in a patterned input environment regardless of the initial condition.

EXPERIMENTAL VALIDATION OF THEORETICAL ASSUMPTIONS

Learning rules such as BCM are instructive because they allow one to connect the elementary principles of synaptic plasticity with the systematics of learning and memory. However, showing that a learning rule yields desirable properties in artificial neural-network models is no proof that it is actually implemented by synapses in the brain. In general, there are two ways to test and/or distinguish between theories. One is to compare the predicted consequences of a theory with experimental findings; in the case of the BCM theory this comparison has yielded impressive results. The other approach is to verify experimentally the underlying assumptions. Gains in our understanding of the mechanisms of synaptic plasticity evolved to a point, about ten years ago, where such experiments became possible. The biological feasibility of the BCM algorithm quickly became apparent.

The search began in the hippocampus, an ancient structure lying deep inside the brain. The hippocampus, because of its cellular architecture and arrangement of axonal connections, is a favorable preparation to study the mechanisms of synaptic transmission in the central nervous system. Four key discoveries laid the foundation for the experimental validation of the theory. First, Bliss and Lømo showed in 1973 that synapses in the hippocampus could be potentiated (strengthened) when stimulated electrically at high frequencies, and that the synaptic change was very long lasting (it thus came to be known as long-term potentiation, or LTP). Second, in 1986 a number of groups reported that LTP is a specific consequence of the coincident release of the neurotransmitter glutamate from presynaptic axon terminals and the strong depolarization of the postsynaptic cell membrane (high-frequency stimulation induces LTP because it satisfies these conditions). Third, it was discovered that the coincidence of glutamate and postsynaptic depolarization caused a specific neurotransmitter receptor, the NMDA receptor, to open its ion channel and allow calcium to enter the postsynaptic neuron. Finally, it was shown that NMDA-receptor activation and postsynaptic calcium entry were required for the induction of LTP in hippocampus. These findings (and others) suggest that the entry of calcium through the NMDA receptor triggers the biochemical changes in the postsynaptic neuron that cause the synapse to potentiate.

The potentiation of synapses that are active at the same time as strong postsynaptic activation is an assumption of most "Hebbian" learning rules, including BCM. Thus, in 1987 we suggested that the modification threshold of the BCM theory, θ_m, corresponds to the level of postsynaptic activation at which a critical level of calcium passes through the NMDA-receptor channel. Associating θ_m with a critical level of NMDA-receptor activation and postsynaptic calcium entry led to the additional hypothesis that input activity, which fails to activate NMDA receptors beyond the level required to trigger LTP, should cause long-term depression (LTD) of the active synapses. We initially turned to the hippocampus to test this hypothesis.

High-frequency stimulation is particularly effective in inducing LTP because it produces a strong postsynaptic response. In an effort to provide a high level of presynaptic input activity without producing a postsynaptic response so large that it yielded LTP, Dudek and Bear tried extended periods of low-frequency stimulation in the range of 0.5–10 Hz.[3] They discovered that stimulation of the synapses at frequencies between 0.5 and 3 Hz, which produces only a weak postsynaptic response, did indeed trigger LTD.

By varying the stimulation frequency but holding the stimulation intensity and the total number of pulses constant, it is possible to derive a function relating synaptic plasticity to stimulation frequency. Little or no plasticity is observed at frequencies less than 0.1 Hz, robust LTD is observed using 1 Hz stimulation, and LTP is observed using stimulation frequencies greater than 10 Hz (figure 8, plate 10). Interestingly, 10 Hz stimulation produces, on average, neither LTD nor LTP. Because the amount of presynaptic activity is held constant (900 pulses) in these experiments, the frequency-response function in figure 8 is equivalent to the BCM modification function in figure 7.

Next we wished to examine the generality of this type of synaptic plasticity. It has been appreciated for some time that the neocortex on the surface of the cerebrum is a major site of memory storage. Recent work at Brown University and elsewhere has demonstrated LTP and LTD in the neocortex and has shown that these forms of plasticity are governed by mechanisms similar to those described in the hippocampus. Of particular relevance are data from a group at Yale University, led by Anne Williamson, showing that the same principles of synaptic plasticity apply in the human inferotemporal cortex, the region implicated as a repository of visual memories (figure 9, plate 10). Together, the data support the idea that very similar principles guide synaptic plasticity in widely different regions of the brain.

Despite this success, a critical unproven assumption of the BCM theory remained. To account for learning, the modification threshold, θ_m, must vary according to the history of postsynaptic cortical activity. Kirkwood, Rioult, and Bear approached this question by comparing the frequency-response function in the visual neocortex of normal animals with that in the visual cortex of animals reared in complete darkness (the visual cortex is that part of the neocortex that receives information specifically from the eyes).[4] In accordance with theoretical predictions, they found in the visual cortex of light-deprived rats that LTP is enhanced, and LTD is diminished, over a range of stimulation frequencies. These findings support the concept that θ_m is set according to the activation history of the cortex.

While the investigation is far from complete, enough has been done to illustrate how a theoretical "learning rule" has helped guide experiments into the elementary mechanisms of synaptic plasticity in the brain. The data clearly indicate that the key assumptions of the BCM theory have a biological basis in the hippocampus and neocortex. Active synapses can undergo LTD as well as LTP, the key variable determining the sign of the modification is the level of postsynaptic response, and the depression-potentiation crossover point θ_m varies depending on the history of cortical activity. Work is now underway to understand the detailed molecular mechanisms of LTP and LTD, as well as their regulation by experience.

In addition to serving as a guide for experiments, theoretical analysis can also be used as a bridge to connect these mechanisms with their consequences. The BCM theory shows that a natural consequence of the mechanisms of LTP and LTD can be the experience-dependent modification of stimulus-selective cellular responses leading to the distributed storage of information in the brain. We therefore have the exciting possibility that the BCM theory provides a mathematical structure that can link the detailed molecular mechanisms of long-term synaptic plasticity to the systems-level property of learning and memory storage by the brain.

In the field of learning and memory, it is reasonable to say not only that more progress has been made than is generally appreciated but that even we are astonished, given the level of skepticism displayed when theories of memory based on synaptic modification were first introduced twenty years ago. The field has changed. It is no longer uncommon to see experimental papers with lengthy discussions of theoretical issues. Nor is it rare to have new theories proposed that incorporate detailed biochemical mechanisms for synaptic plasticity. Theoretical structures have and will continue to provide illumination of the neurobiological basis for learning and memory storage.

A THEORY OF MIND?

Studies of the brain will give a more detailed understanding of the genetic, molecular, and cellular basis of various brain functions, including learning, memory storage, and information transfer. Imaging studies will help us to understand what parts of the brain are most active in processing various types of information. How spatial and temporal information is put together and how the various brain subsystems interact to enable us to recognize, associate, and reason—only dimly understood at present—will most likely be elucidated in the not-too-distant future.

But we are not yet ready to uncork the bottles of champagne; for how all of this neural machinery is put together to produce our self-awareness, our consciousness, is not understood. This problem seems so baffling that first we must convince ourselves that a solution is possible. Think about it in this way: Is it possible for us to construct, from ordinary materials such as synapses, neurons, and systems of neurons, a machine that is conscious? The difficulty of this question is such that we have been subjected to many evasions—Cartesian dualisms (the mind is distinct from the brain), denial of the phenomena (consciousness and feeling are "epiphenomena" [whatever those are] and do not have to be explained or, in the extreme, do not exist), solutions of one mystery by invoking another (e.g., consciousness arises in the quantum measurement process or where gravity meets quantum theory), refusal to confront the issue (e.g., consciousness arises "somehow" when a machine executes the proper algorithmic processes), retreat under the cover of positivist philosophy (how would we know if a machine were conscious?), and so on.

But we have heard such arguments before. They seem to be typical responses to the frustration of failure when attacking really difficult scientific problems. First, try and fail; follow this by proving that a solution is impossible or irrelevant. Toy with the notion that a new law of nature is involved. Then, when the solution is found, complain that it is really trivial or (even better) that it was suggested in some obscure comment once made in a paper published by the complainer a long time ago.

What we must understand is how consciousness arises as a property of a very complex physical system. The scientific problem, as we see it, is to construct from material components such as neurons and systems of neurons the simplest entity that performs the most primitive conscious act. (To paraphrase Bourbaki, a beautiful problem: one that can be stated very simply and will, no doubt, have a very complex solution.)

In considering this question, we must overcome the fear of, or aversion to, assumptions about the internal workings of the mind that cannot be directly verified by experience. Successful science has given us just such "machines" (the greatest include Newton's laws, Maxwell's equations, and Schrödinger's equation). Such entities as molecules or atoms were assumed to exist (an assumption that was vigorously contested in the nineteenth century with positivistic-type arguments) long before they were "seen." It is not necessarily the case that every element of the "behind the scenes" machinery can be directly observed. In quantum mechanics, for example, the wave function is not directly observable. The consequences of this sometimes invisible machinery can, however, be put into correspondence with experience.

A satisfactory theory of mind not only allows but, in our opinion, requires the introduction of mental entities constructed from the materials available. We will be satisfied only when we see before us constructs that can have mental experience, when we see how they work and how they come about from more primitive entities such as neurons. It is possible that there is no sharp demarcation between consciousness and nonconsciousness (just as we would say today that no such sharp distinction exists between the categories organic and inorganic, plant and animal, or even living and nonliving).

Of course, it could turn out that we must invoke a new "law of nature"—follow Descartes and pour the conscious substance into the machine. But the conservative scientific position is to attempt to construct this seemingly new and surely very subtle property from the materials available—those given to us by physicists, chemists, and biologists (as has been done many times before: celestial from earthly material, organic from inorganic substances, the concept of temperature from the motions of molecules, or light from electricity and magnetism). The unrepentant reductionist believes that this construction can and will be made; that it will in no way diminish the value or significance of what has been constructed; that, to paraphrase Santayana, "All our sorrow is real, but the atoms of which we are made are indifferent."

Success would no doubt be magnificent but failure might be more so. If we cannot perform the reduction, then we will genuinely have made one of the most profound discoveries in the history of thought—the consequences of which would shape our conception of ourselves in the deepest way.

ACKNOWLEDGMENTS

The work on which this essay is based was supported by the Charles A. Dana Foundation, the National Science Foundation, the Office of Naval Research, and the National Eye Institute.

ENDNOTES

[1]Leon N Cooper, Fishel Liberman, and Erkki Oja, "A Theory for the Acquisition and Loss of Neuron Specificity in the Visual Cortex," *Biological Cybernetics* 33 (1979): 9–28.

[2]Elie L. Bienenstock, Leon N Cooper, and Paul W. Munro, "Theory for the Development of Neuron Selectivity: Orientation Specificity and Binocular Interaction in the Visual Cortex," *Journal of Neuroscience* 2 (1982): 32–48.

[3]Serena M. Dudek and Mark F. Bear, "Homosynaptic Long-Term Depression in Area CA1 of the Hippocampus and the Effects of N-methyl-D-aspartate Receptor Blockade," *Proceedings of the National Academy of Science* 89 (1992): 4363–4367.

[4]Alfredo Kirkwood, Marc G. Rioult, and Mark F. Bear, "Experience-Dependent Modification of Synaptic Plasticity in the Visual Cortex," *Nature* 381 (1996): 526–528.

ADDITIONAL REFERENCES

Bear, Mark F. "A Synaptic Basis for Memory Storage in the Cerebral Cortex." *Proceedings of the National Academy of Science* 93 (1996): 13453–13459.

Bliss, Timothy V. P., and Graham L. Collingridge. "A Synaptic Model of Memory: Long-Term Potentiation in the Hippocampus." *Nature* 361 (1993): 31–39.

Chen, Wei R., et al. "Long-Term Modifications of Synaptic Efficacy in the Human Inferior and Middle Temporal Cortex." *Proceedings of the National Academy of Science* 93 (1996): 8011–8015.

Rolls, Edmund T., et al. "The Effect of Learning on the Face-Selective Responses of Neurons in the Cortex in the Superior Temporal Sulcus of the Monkey." *Experimental Brain Research* 76 (1989): 153–164.

Jean-Pierre Changeux

Drug Use and Abuse

Le seul plaisir durable est dans la connaissance objective.
—D. Diderot

THE USE OF DRUGS TO AFFECT conscious states in humans goes back almost to the origins of humanity. The pollen of eight medicinal plants was discovered deposited intentionally in a 60,000-year-old tomb in Iraq. The shelves of the Assurbanipal library held tablets in cuneiform writing that listed some 250 plants, 120 mineral substances, and 180 animal substances. Opium poppies and cannabis were mentioned together with Mandragora, *Ricinus*, *Elleborus*, incense gum, and myrrh. Indeed, shamans still exploit the active principle of plants to induce ecstatic states through which they communicate with spirits, devils, and deities and acquire from them the "knowledge" to heal and restore harmony in the social group. It is interesting to note that from the earliest examples we have of the use of these substances, strong links between drugs, medicine, and social order have been established. According to Barbara Meyerhoff, in their pilgrimage to Wirikuta the Huichol *peyoteros* enter a world of dreams where social distinctions are erased, the natural and supernatural orders have been overcome, and time itself is obliterated.[1] In their mystical experience, even the ego is lost. Furthermore, in occidental societies, religious rituals have made extensive use of drugs. Alcohol was the basic stimulant of the Bacchus cult in ancient Greece. Both the Old and New Testaments repeatedly refer to the use of wine, from Noah's and

---

Jean-Pierre Changeux is professor at the Collège de France and the Institut Pasteur.

Lot's drunkness to the wedding at Cana and the Last Supper. Incense burning is still an active component of many religious rituals. In fact, ritualization became the first efficient procedure to control the use of drugs in human societies; another emerged with the rise of rational medicine in Greece.

Initially, the word *pharmacon* referred to both the magic action of plants to heal (or to poison) and to the "demon" causing the disease in the body. Secrecy and mystery were progressively replaced by the separation of secular physicians from the caste of priests, the spread of Hippocratic medicine, and the occurrence of open public debates between different schools of thought, which together resulted in the emergence of rational medicine. As a consequence, diagnoses were made and remedies proposed on the basis of the "active principles" of plants that were later defined as chemical substances, and thus pharmacological agents. Medicine achieved a powerful and socially beneficial way to regulate the use of drugs.

In Western society, recreational use of the pharmacologically active derivatives of plants, such as tobacco or coffee, became popular during the sixteenth century following the return of European explorers. Yet throughout the nineteenth century, the entry to "artificial paradises"—for instance, through hashish or opium smoking—was limited to a select few socialites, poets, and artists. Quite surprisingly, with the invention of the syringe and hypodermic needle, a number of medical doctors also became recreational users of mind-altering drugs. Meanwhile, in his *L'Abattoir*, Emile Zola linked the compulsive use of alcohol with poverty and social distress. *L'Abattoir* was a dramatic literary evocation of the drug abuse that plagues Western society today, causing major health hazards and world-wide trafficking.

As drug abuse expanded, repressive laws flourished without regard its social origins. Nineteenth-century laws were aimed at preventing criminal use of poisons (as opposed to pharmaceutical use), and thus the consumption of substances with euphoric effects was not repressed. The use of opiates in society first became an offense at the time of World War I, but it was only in the 1950s that the distinction between the patient and the user was formally established. The concept that judicial courts could force, or at least propose, therapy as an alternative to imprisonment for drug addicts was introduced into law only in 1970 (in France), albeit with limited success. Despite strongly repressive laws throughout the world (about four hundred thousand individuals are currently incarcerated in the United States for violation of drug laws) and many costly "anti-drug wars," drug abuse remains an alarming health hazard throughout the world. Illicit drugs account for 35 percent of new cases of AIDS in the United States. Heroin-related deaths rose in U.S. metropolitan areas from thirteen hundred in 1985 to thirty-five hundred in 1994. The probability of a cocaine or heroin seller being incarcerated has risen sharply since about 1985, but that has not led to reduced availability, and prices remain constant.[2]

The aim of this essay is to shed light on the knowledge gained in recent years concerning the neurobiology and pharmacology of drug action and addiction. It will demonstrate that the distinction made between licit and illicit drugs in present legislation is not based upon any coherent scientific thinking, and that drug use and drug abuse are *not* synonymous. Although the social and economic aspects of drug consumption have had a dominant impact in the recent evolution of drug abuse, they will not be dealt with. The present discussion of the scientific side of the problem should, by no means, divert attention from these social issues, which are often deliberately underevaluated and insufficiently tackled on a global scale in an open-minded, objective, and concerted manner.

DRUGS AS ANALOGUES OF BRAIN NEUROTRANSMITTERS

Our brain is made up of a very large number of nerve cells or neurons (up to one hundred billion), which are themselves limited by multiple connections (about ten thousand per neuron on average) established through their axonal and dendritic processes. The complexity of brain connectivity is astonishing. Yet at the subcellular and molecular scale, the picture simplifies. While Ramon y Cajal initially proposed that neurons are independent and contiguous units, electron microscopy showed that at the level of the contacts between neurons, or *synapses*, the cell membranes do not fuse but are separated by a significant gap. The distances between the cell surface and the conductance of the external medium in the majority of neurons are such that electrical communication cannot take place. Chemical substances, referred to as *neurotransmitters*, thus relay the electrical impulses to bridge the gap.

When the electrical impulse invades the nerve terminal, the neurotransmitter stored in the ending is released into the cleft as a brief (about one millisecond) high-concentration chemical pulse that diffuses rapidly through the intercellular space until it reaches the next cell membrane. There, the neurotransmitter elicits a response in the form of either a flow of charges across the membrane (thus generating electrical currents) or a set of intracellular reactions (without necessarily being accompanied by a change of electrical properties). In some instances, the distance between the release site and the target of the neurotransmitter becomes relatively large—on a scale of millimeters or even centimeters. To some extent, the mode of communication then resembles that of hormones. In all these instances, the neurotransmitter serves as a critical chemical signal in the communication between nerve cells (figure 1, plate 2).[3]

About fifty neurotransmitters have now been identified in our brains. Most of them are small organic molecules with a molecular weight of about 100 to 200, but the list can be extended to the many neuro-peptides that, like our body proteins and many hormones, are linear chains of amino acids having variable length. These diverse neurotransmitting molecules may exert strikingly different effects on the firing state of their target neurons. Some, like glutamate or acetylcholine, stimulate firing and are referred to as *excitatory*; others, like γ-aminobutyric acid (GABA), decrease firing and are called *inhibitory*. Moreover, the same neuron may synthesize and liberate up to ten chemical messengers. The diversity of chemical coding thus accessible to nerve cells is thus very large.

In midbrain sections, clusters of cell bodies discretely stain for a particular group of neurotransmitters that play a crucial role in drug addiction: dopamine, norepinephrine, acetylcholine and serotonin. Despite being relatively few in number (on the order of a few thousand neurons), these neurons exert global effects on large regions of the brain (for some of them over the entire brain) through the innumerable, widely divergent ramifications of their axons and the "volume" release of the neurotransmitter they synthesize.

Within this general context of chemical neurotransmission, a simple principle emerges for drug action in the brain. As we shall see, drugs may be viewed as *chemicals mimicking—or interrupting the effect of—endogeneous neurotransmitters in brain communications* (figure 10, plate 11).[4]

Opium contains several active organic molecules, in particular *morphine*, from which many active analogues have been synthesized; these include heroin (which behaves like morphine) and naloxone (which acts as an antagonist of morphine). *Methadone*, currently used as a substitute for morphine, bears no obvious chemical relationship to morphine yet was designed on the basis of the three-dimensional structure of the molecule of morphine, and indeed assumes a similar configuration in solution. What principally distinguishes it from morphine is that methadone can be taken orally, without requiring hypodermic needles, thus reducing the risk of infection with either the AIDS or hepatitis virus. The brain does not synthesize morphine but rather a set of endogeneous peptides, the opioid peptides, which include the five amino-acid enkephalins and the much-longer endorphins and dynorphin. Structural analogies between the folded peptide and the alkaloid molecules have been thought to explain their similar pharmacological action—analgesia and euphoria, as well as respiratory depression—thereby explaining the danger of overdoses. Opioid drugs, therefore, behave as functional and structural analogs of the endogeneous neurotransmitter peptides, including enkephalins, endorphins, and dynorphin.

Other drugs of important therapeutic use are the *barbiturates* and the *benzodiazepines*. Both are synthetic compounds. The former were synthesized in 1864 (on the feast day of St. Barbara), the latter only recently and accidentally. Both of them, but especially the benzodiazepines, reduce anxiety and aggressivity. They give a feeling of sedation, relax muscles, and induce sleep. Quite remarkably, they have become some of the most widely prescribed drugs in the pharmacopoeia despite the fact that they produce strong dependence and abuse. They augment the action of the inhibitory neurotransmitter GABA on its target neurons, even without showing evident structural analogies with it.

Nicotine and *cannabis* have been primarily used as recreational drugs in Western societies for centuries. There are increasing indications that nicotine and its derivatives act as cognitive enhancers and that cannabis functions as a painkiller in severe diseases. Nicotine is unambigously both a structural and a functional analogue of acetylcholine. The active ingredient in cannabis is 9 tetrahydrocannabinol; this molecule shows structural resemblance to anandamide (from the Sanskrit *ananda*, "felicity"), a recently discovered endogeneous molecule that functions in brain communications.

Despite intense research, the actual mechanism (or perhaps mechanisms) of the action of *alcohol* as a drug is not fully elucidated. Its analogy with defined endogeneous brain substances remains a challenging question. Several possibilities have been raised, none of which are exclusive. It could possibly act as an antagonist of the excitatory neurotransmitter glutamate, or perhaps as an enhancer of the inhibitory neurotransmitter GABA, along the lines of the benzodiazepines.

Cocaine, present in the leaves of a South American shrub (and other synthetic molecules, called *amphetamines*) has potent psychostimulant action, and inducing alertness and euphoria. Cocaine is traditionally used by high-altitude populations in the Andes to improve their ability to work without excessive fatigue. These stimulants generally enhance sex drive; intravenous amphetamine injection can be so intense as to be described as "orgasmic." Yet the clinical applications of amphetamines and cocaine are very limited. Clearly, these molecules are structurally related to dopamine and norepinephrine. Their common general mechanism is to increase the level of these neurotransmitters in interneuronal spaces.

Lastly, the *hallucinogenic* or *psychedelic* drugs have fascinated writers and artists since Aldous Huxley's time because of the hallucinatory mental states they create. Mescaline is the active molecule of the peyote cactus that the Huichols eat in their trip to Wirikuta. Hoffman, who synthesized lysergic acid diethylamide (LSD), discovered its hallucinogenic action by chance. Both mescaline and LSD are nonaddictive, yet the occasional "bad trip" they cause can lead to persistent mental disorders. LSD action appears related to that of serotonin in the brain; mescaline, to that of norepinephrine.

Most known drugs thus mimic or enhance the action of endogeneous chemical messengers already present in the brain. They may be viewed as cultural representations of our brain chemistry, patiently selected by generations of human beings not only for their medicinal usefulness but also for their "ineffable" and "mystical" actions on the human psyche. The shift from socially regulated use to the current (distressing) abuse is a consequence of the dramatic reshuffling of our social world, rather than of any significant change of our brain chemistry. (For an illustration of the similarities between neurotransmitters and drugs of abuse, see figure 11, plate 12.)

NEUROTRANSMITTER RECEPTORS AS THE
PRINCIPAL TARGETS FOR DRUGS OF ABUSE

Little has yet been said about the precise molecular targets of drugs in the brain. Where do they act? The issue was raised by Claude Bernard in his "Leçons sur les substances toxiques et médicamenteuses," delivered in 1857 at the Collège de France, with the aim of localizing the effect of curare, the toxic substance of the poisoned arrows of American Indians. He described an ingenious series of experiments that allowed him to demonstrate that "curare blocks the communication between peripheral motor fibers and skeletal muscle," thus serving as a "chemical lancet" to dissect the nervous system.[5] Inspired by Bernard's experiments, the British pharmacologist John Newport Langley's further analyses in 1905 of the action of nicotine on the muscle of the fowl showed that curare antagonizes the excitatory action of nicotine; he concluded that a "receptive substance," today called a *receptor*, "combines with nicotine and curare" and "is not identical to the substance which contracts."[6] Accordingly, nicotine would bind to the receptor, where it would mimic the endogenous neurotransmitter acetylcholine and cause an electrical response of the membrane. This action results from the opening of a channel allowing positively charged sodium and potassium ions to flow through the membrane, thus creating an electrical current. Curare, in contrast, would act at this level as an antagonist of nicotine or acetylcholine.

Although the receptor concept was quite elegant, there were no tools to isolate and identify chemically the molecules present in such minute quantities. Consequently, the scientific community remained skeptical. Moreover, from a conceptual point of view, physiologists and pharmacologists had trouble conceiving how the binding of a chemical to the cell surface would create the "punctures" that were understood to permit ions to flow through the membrane. A plausible though hypothetical answer came from biophysical and structural studies on specialized proteins (enzymes from bacteria or oxygen-carriers, like hemoglobin in vertebrates) that behave as molecular switches. These proteins were named *allosteric* because they carry at least two distinct categories of binding sites with different "steric" specificities: one for the regulatory signal, the other for the substrate of the biological activity.[7] In the case of neurotransmitter receptors, the first one would be the drug-neurotransmitter binding site; the second, the ion channel. The coupling between them would be a discrete and reversible transition between two conformations of the molecule, one with the channel open and the other with the channel shut. The physiological signal (acetylcholine) or the drug (nicotine) would then "open the lock"; curare would close it.

To evaluate the merit of these ideas and progress further in the understanding of drug action, the chemical identification of a neurotransmitting drug receptor became a priority. In the early 1970s, it happened with the receptor for nicotine following a rather unchartered path. First, the biochemist had to start with a tissue rich in synaptic material and thus expected to contain significant amounts of receptor. The electric organs of electric fish—the freshwater Amazonian electric eel and the torpedofish from the French oceanic coast—were recognized by David Nachmansohn in the 1930s (then working in France) to offer an adequate system. These tissues indeed contain billions of chemically identical acetylcholine synapses with properties very similar to those of the junction between motor neuron and muscle. Such a gigantic accumulation of synapses creates a rare amplification that is very useful for the chemist. The quantities of receptor protein available reach several grams per electric organ.

Still, the separation of the receptor from crude extracts of electric organ was needed. Indeed, many different sites and molecules were anticipated to bind the neurotransmitter in the electric organ "soup." One toxin characterized by Chen-Yuan Lee of Taiwan University—from the venom of poisonous snakes (cobra or bungarum)—happened to be an adequate flag, since it labels the synaptic site where acetylcholine binds with an extreme selectivity and poor reversibility. The joint use of the fish electric organ and snake venom toxin, initially done in my laboratory, was indeed a decisive step in the identification of the acetylcholine nicotinic receptor. Moreover, to identify the lock and to describe both how it is built and how it works, one needed to have it in solution without the key blocked in it. After a long and difficult quest, this was achieved by releasing the receptor in solution from the synaptic membranes with a mild detergent that preserves the functional state of the molecule. The receptor molecule could subsequently be purified with its binding site intact and functional, using the toxin only as an assay, and its biochemical identification as a single molecular entity could then be achieved.[8]

The receptor for nicotine and curare in mild detergent solution is a large protein, which (as proposed initially by John Langley) does not show any structural relationship with the contractile equipment of the muscle cell. Its molecular weight is 300,000, or over four times that of hemoglobin. It results from the assembly of five subunits that each cross the membrane several (four) times (figure 12, plate 13). The purified protein suffices to mediate the physiological response to the neurotransmitter and to the drug nicotine. Indeed, it can be reinserted into an artificial system made up of a bi-layer of lipids, yielding a fully functional system. The reconstituted receptor responds to acetylcholine or nicotine by the opening of the selective ion channel. In other words, this single molecular species suffices to convert the acetylcholine signal into an electrical response, and the drug nicotine mimics the action of the physiological neurotransmitter on this receptor protein.

Subsequently, recombinant DNA technologies came into the picture, leading to the identification of the gene coding for the subunits of the electric organ receptor. The late Shosaku Numa, who pioneered these studies in Kyoto, further showed that human skeletal muscle receptors are very similar to those of fish. Muscle gene fragments were then used by James Patrick and Steven Heinemann from the Salk Institute as "hooks" to "fish out" brain genes. Ten new brain subunits were identified and found to associate into functional molecules. After all, we have in our brain drug receptor molecules that are, from an evolutionary point of view, as old or even older than the electric fish. Their birth date is almost a billion years ago.

Chemical investigations now reveal the fine structure of the drug target site at nearly the atomic level. The five subunits that compose the molecule may differ. Two of them, called *alphas*, were shown by Arthur Karlin from Columbia University to bear the primary responsibility for recognizing acetylcholine or nicotine. Recent work indicates that the drug binding pocket in fact lies at the boundary between an alpha and a nonalpha subunit. There, nicotine, acetylcholine, or even the antagonist curare are firmly captured by at least five (and possibly six) "fingers" that, as anticipated by Paul Ehrlich at the end of the nineteenth century, establish weak though multiple chemical bonds with the drug.

Using a "reactive cork" that irreversibly plugs the ion channel, the walls of the ion pore were shown to occupy a distinct part of the receptor molecule. The distance between the acetylcholine binding sites and the ion channel measured with optical "rulers" is quite large on the scale of the receptor molecule—in the range of one fourth to one third of its diameter. Moreover, structural investigations and electrophysiological recordings in reconstituted systems strongly support the view of a reversible conformational or "allosteric" switch between a silent resting state and an active open channel state, which are spontaneously accessible to the molecule and stabilized by both acetylcholine and nicotine. In other words, the endogeneous brain messenger and the drug have similar effects at the molecular level as regards the properties of the receptor lock.[9]

Receptors for other drugs and neurotransmitters look rather similar, even those that display an "inhibitory" action responding in a direction opposite to that of nicotine. Instead of facilitating the transport of positively charged sodium and potassium ions, their channel facilitates that of negatively charged chloride ions. Only a few atoms in their channel domain are needed to determine the difference, thus resulting in the inhibitory versus excitatory character of the response. These inhibitory receptors we would therefore expect to decrease the level of excitability; that is to say, they keep people "quiet." Indeed, the tranquilizers, like the benzodiazepines mentioned earlier, potentiate the action of some of these inhibitory receptors, yet at a site distinct from the neurotransmitter binding site. Binding of the drug to this site facilitates the switch of the receptor lock in favor of its open state. This is, to some extent, analogous to the sort of complex lock on a safe that requires two keys to be turned simultaneously to open the door. This "allosteric" interaction of a new kind thus solves the paradox raised by the benzodiazepines having a common effect on the receptor despite structural differences between the "artificial" drug and the "natural" neurotransmitter.

As alternatives to enhancing the flow of ions through channels, other drugs and neurotransmitters may interact with a different family of receptor molecules that instead modulate intracellular chemical signalling. This family includes the receptors for many drugs of abuse (such as morphine and 9 tetrahydrocannabinol), but also for the neurotransmitters norepinephrine and dopamine, for odorants, and even for single photons (in the case of the eye pigment rhodopsin). These receptors—of which there may be more than a thousand distinct forms in our body—are also embedded in the cell membrane but share a common molecular organization. Their amino-acid chain crosses the membrane seven times, forming a barrel-like structure with a pocket where the drug molecule binds.[10] As in the case of the channel-linked receptors, binding of the drug or neurotransmitter at this level stabilizes a conformation of the receptor, which subsequently triggers a cascade of molecular processes on its intracellular face via a set of specialized regulatory proteins called G-proteins (figure 13, plate 13). These molecules take the relay and transmit signals to a wide variety of intracellular regulatory pathways.

Special mention should be made of cocaine and amphetamines, which, rather than acting on neurotransmitter receptors, interfere with their presynaptic management (figure 10, plate 11).

Cocaine blocks the transport protein involved in the reuptake of the once-released norepinephrine or dopamine, whereas amphetamine incorporates as a "false neurotransmitter" in the nerve ending and, once there, enhances the release of the endogeneous neurotransmitter. In both cases, the net result is an increase in the local concentration of the neurotransmitter. As a consequence, both drugs of abuse potentiate the action of the endogeneous neurotransmitters on their receptors. This indirect action reinforces the general conclusion that drugs mimic or enhance (and sometimes antagonize) the effect of brain neurotransmitters on their receptors. This is equally true for nicotine, opiates, benzodiazepines, or cocaine; in other words, it is the same for illicit as well as for licit drugs. Thus, at the level of their molecular target, the distinction between licit and illicit drugs does not rely on any coherent scientific principle.

Obviously, this does not mean that all drugs are equal in terms of the morbidity and mortality risks caused by their intake. But here again, no obvious correlation seems to exist between the short- or long-term toxicity of psychoactive drugs and their legal status. Risks vary with accompanying substances and/or infectious agents, but primarily with dosage and with individuals. As mentioned, many of them—in low doses, but sometimes also at higher doses—are beneficial and serve as efficient medications. The frequency of use is also a factor affecting risk to the user. Addiction may indeed take place.

DRUG ADDICTION

The recreational and pleasurable use of drugs, associated at least in some cases with euphoria and/or flashing hallucinations, would generally be of limited risk if it did not produce compulsive self-administration. In the early stages of drug use, the *user* is most often viewed as having free choice, being fully capable of resisting any attractive property of the drug. However, progressively the user loses control over drug intake; he becomes a drug *abuser* and, finally, drug-*dependent*. He then needs the drug to function within normal limits. He seeks higher doses of the drug to achieve the same effect—that is, a degree of *tolerance* develops. Moreover, cessation of administration leads to a with-

drawal syndrome characterized by serious physiological disturbances. For instance, morphine addicts show symptoms somewhat resembling severe influenza with yawning, pupillary dila-
tation, fever, sweating, piloerection (thus the description "cold
turkey"), nausea, diarrhea, and insomnia. Patients are extremely
restless and distressed, and have a strong craving for the drug.
The symptoms progressively disappear after a week or two, but
residual physiological and psychological abnormalities persist
for many more weeks. Administration of morphine rapidly abolishes the abstinence syndrome.

The term "addiction" is frequently restricted to the extreme or
psychopathological state where control over drug use is lost.
Physical withdrawal syndromes caused by various drugs differ
in substantial ways. However, subjective symptoms including
dysphoria, depression, and anxiety are common to withdrawal
from several drugs, such as sedative-hypnotics, nicotine, opiates,
and psychomotor stimulants. Physical dependence has consequently often been contrasted with a so-called psychological
dependence related to drug-seeking behavior and compulsive
drug consumption. But recent research has shifted the emphasis
toward two aspects of addiction. First, compulsive self-administration of drugs does not dominate behavior simply to alleviate
aversive withdrawal symptoms. On the contrary, habit-forming
properties of drugs become increasingly linked to effects on
brain circuits that are involved in the service of natural *reward*
(or positive reinforcement). Second, drug dependence and withdrawal phenomena are not directly linked with alterations of the
autonomic peripheral nervous system that causes the withdrawal
symptoms. Rather, they result from *adaptations within* specialized reward (or reinforcement) circuits in the brain that most
drugs of abuse activate in common.[11]

Among the philosophers of ancient Greece, Epicurus stands
out for defending a rather simple, though challenging, thesis:
human beings are pulled forward toward and by nature seek
pleasure, whereas they flee from and reject pain. A naive neural
model for the Epicurean position would thus be that we have in
our brain "pleasure centers" and that we constantly search out
situations that sooner or later lead to their stimulation. The
situation is not that simple. Yet since Pavlov, experimental psy-

chology—and studies of learning in particular—have placed considerable emphasis on reward (or punishment) to validate and reinforce (or eliminate and repress) a defined behavior. For example, in these standard tasks, the reward is delivered as a palatable food, the punishment as an electric shock. In the late 1950s, Olds and Milner began to look for a localized system that served to mediate these effects in the brain.[12] They implanted chronic stimulation electrodes in different points of the brains of rats and left each rat freely moving in a Skinner learning box. This differs from standard operant conditioning in that when it steps on the pedal, instead of getting a food pellet, the rat receives an electric shock to its own brain by means of the implanted electrodes.

In the absence of any reward, the rat steps on the pedal about twenty-five times during the first hour. If the pedal pressing causes a reward—for instance, when an electrode is implanted in a brain region called the septum—the rate rises to two hundred or more per hour, even during the first hour. If the pedal-pressing produces punishment at the point where the electrode is located, the rate drops radically to one or two responses per hour. Electrical self-stimulation rates may become as high as seven hundred per hour when the electrode is implanted in an area called the interpeduncular nucleus. They tend to decrease when the electrode is placed forward in the anterior brain. The rat may engage in self-stimulation for more than twenty-four consecutive hours, then sleep for a few hours and resume self-stimulation at the same rate. Hunger does not distract from self-stimulation. Olds and Milner made clear that, distinct from the cell group that mediates such *primary* rewarding effects, other subsystems mediate the specific drives for hunger and for sex. They also noted that such subsystems of rewarding (or punishing) structures may be sensitive to different chemicals. Subsequent experiments indeed showed that in the presence of nicotine and cocaine the rat begins to self-stimulate at electrical stimuli that were previously ineffective. In other words, these drugs facilitate electrical self-stimulation. Even closer to the human compulsive use of drugs, methods have now been established to enable freely moving animals (rats, mice, and even monkeys) to earn intravenous drug injections. Self-administration of opiates, cocaine, amphetamine, or nicotine in animals thus results in forming a drug habit. No animal models incorporate all the elements of addiction in humans; yet the human addict, by analogy, would become trapped in a drug self-administration process from which escape cannot take place without extreme difficulties.

According to Koob and Le Moal, most human drug users do not become drug abusers or drug-dependent, but a combination of multiple factors such as availability, genetics, history of drug use, stress, and life events may contribute, in some *individuals*, to the transition from drug use to drug addiction.[13] Addiction would then develop as a dysregulation in the reward processes that contribute to the equilibrium of the organism with its environment using its own physiological and cognitive or behavioral capabilities. Failure to self-regulate drug use resembles to some extent pathological gambling, binge eating, compulsive exercise, compulsive sex, mystical religiosity, and others. Similar patterns of spiralling distress-addiction cycles may start by a first self-regulation failure that sets up a cycle of repeated failures to self-regulate in which each violation brings additional negative affect. Animal studies show that stresslike stimuli reinstate self-administration that has been previously extinguished. Genetic vulnerability to drinking alcohol may differ between rodent strains. In addition, vulnerability to drug taking may be influenced by a history of drug experience. The combination of genetic and environmental factors can dramatically change the response to drugs in animal systems. In humans, similar complex combinations are anticipated, with the additional complexity of the emotional components of social life and personal experience.

It is important to understand that the primary reason why addicts seek drugs is for the positive reinforcement by the drug, usually associated with pleasure or euphoria. Drugs act as artificial reinforcers of endogeneous neurotransmitters acting on their receptors within privileged reinforcement circuits. What are these neural circuits? In the last decades intense research—primarily by Swedish scientists Arvid Carlsson, Kjell Fuxe, Anders Björklund, and their colleagues—has led to the construction of a central map of positively and negatively reinforcing sites implicating mostly the already mentioned *catecholaminergic* neurons and within this aminergic system, the dopamine *meso-corticolimbic* neurons.[14] As indicated by their name, these neurons have their body in the midbrain (thus "meso"; it is specifically within the *ventral tegmental area*) and project to forebrain structures, including the prefrontal cortex and the "limbic" areas (in particular, a specific ventral subdomain primarily involved in motor behavior and referred to as *nucleus accumbens*). These dopamine neurons differ from the nearby dopamine neurons of the *subtantia nigra* that primarily project to the striatum, the degeneration of which create the movement disorders of Parkinson's disease. The projections to the limbic areas contribute to emotions, hedonic pleasure, and memory; those to the prefrontal cortex to motivation and planning, as well as to the temporal organization of behavior, attention, and social behavior. According to Carlsson, this mesocorticolimbic system would be disturbed in schizophrenia. Strong neurobiological evidence further indicates that this system contributes to the acute reinforcing action of drugs. Together with its projections to output structures, it provides a neural mechanism by which motivation gets translated into action; this mechanism is based on a *common neural circuitry* that mediates the rewarding effect of opiates, nicotine, amphetamines, cocaine, alcohol, and cannabis. (For an illustration of the dopaminergic system, as well as the three other major neurotransmitting systems, see figures 14–17, plates 14 and 15.)

Morphine (or heroin) and nicotine, as well as amphetamine and cocaine-specific receptors, are localized on the dopamine neurons themselves (or on neurons directly connected with them) and from there regulate the release of dopamine. For instance, nicotine strongly enhances the release of dopamine from the mesocorticolimbic neurons. Moreover, many nicotinic receptor subunits are expressed in these dopaminergic neurons, but their respective contributions to nicotine addiction remain enigmatic. Recently, a mouse lacking the gene for a widespread brain nicotinic receptor subunit (referred to as β2) has been constructed. Interestingly, the "knock-out" mouse has lost altogether the high affinity binding of nicotine to the brain, the enhancing effect of nicotine upon dopamine release, and the ability to self-administrate nicotine.[15] Moreover, the electrical activity of the dopamine neurons recorded *in vivo* in monkeys varies with various appetitive stimuli. It correlates remarkably well with positive reward delivery, but also anticipates reward after learning. This activity is the signature of an *adaptation* that accompanies learning.

The dopamine neurons also progressively "adapt" to the repeated administration of drugs. A first sign is the decreased responsiveness of the receptors for the drugs. In his pioneering studies on muscle response to nicotine, Langley already noted that the response amplitude fades—the receptor *desensitizes*—upon repeated applications of the drug. Recent biochemical studies on both electric organ and brain nicotinic receptors have confirmed this observation, and further shown that this adaptive property is built into the molecular structure of the receptor molecule. It may even be viewed as some kind of high-order "allosteric change," much slower than the opening of the ion channel by nicotine or acetylcholine.[16] Analogous structural changes of the receptor-transducing machinery also take place with the seven transmembrane domain receptors, such as the receptors for opioid peptides and drugs. To maintain the response at its initial level, larger doses are thus needed. A first plausible mechanism for tolerance emerges from the desensitization properties of the neurotransmitter receptors and the related signal transduction mechanisms. Yet longer-term adaptations may also take place and would plausibly result from a regulation of the biosynthesis of the receptors themselves and/or of the associated transducing molecules.[17]

In any case, convergent evidence supports the view that at both the behavioral and the neurochemical levels, a *depression of the reward system accompanies drug withdrawal*. Indeed, dopamine depletion systematically accompanies withdrawal from chronic intoxication with cocaine, opiates, or ethanol. A common neurochemical basis may thus be postulated for drug abuse and addiction, based upon a slowly reversible neurochemical "adaptation" of the dopamine reward system. Activation and adaptation of the *dopamine system* thus appears as a *common denominator* for opiates, amphetamines, cocaine, nicotine, ethanol, and possibly cannabis, reinforcing effects and withdrawal consequences after chronic intoxications.[18]

Still much remains to be understood about the neural circuits and the molecular regulatory mechanisms involved in drug action. Yet in the light of knowledge gained in recent years in pharmacology and neurobiology, no coherent scientific conclusion can support the legal distinction between licit and illicit drugs. Drugs that are not prohibited (alcohol, tobacco, pain killers, or neuropsychiatric medications) are potentially just as dangerous from the perspective of potential drug abuse as prohibited drugs. Moreover, toxicity and social cost may be as devastating for several of them as those of prohibited ones.

CONCLUSIONS

Compulsive use of drugs is a universal human fact—a tragic reality of today's society that most governments have failed to control. With the exception of the Netherlands, Western countries assume that the legal prohibition of a limited list of drugs of abuse is an efficient weapon. Meanwhile, drug trafficking on a worldwide scale has developed very profitably, benefitting considerably from repressive laws and clandestine use. According to the Financial Action Task Force, global trafficking earns its perpetrators in the range of $500 billion per annum. In addition, the hidden redistribution of the benefits of these earnings infiltrates the whole of society, with 90 percent of the amount invested in rich countries and only 10 percent in poor countries. The percentage of effectiveness of efforts to combat this money is thought to be less than 1 percent. Faced with the dramatic, and probably deliberate, incapacity of political powers to grasp the issue of drug abuse, the view has been expressed (by such distinguished personalities as Raymond E. Kendall, Secretary General of Interpol, to cite but one), that states should change their antidrug strategies and concentrate efforts on prevention and assistance to drug abusers, rather than on repression. Because of this state of affairs and the alarming growth of certain forms of drug addiction—aggravated in the general context of the AIDS epidemic and the phenomena of social exclusion prevalent in all societies—the French National Consultative Ethics Committee for Health and Life Sciences (CCNE) has for its part chosen to examine drug-related problems irrespective of whether the drugs in question are legal or not. Until now, it appears to be one of the very few such panels, if not the only one, to have faced this issue.[19]

Among its conclusions, the CCNE noted that policies of repression are no longer sufficient to solve problems that involve the use of substances affecting the nervous system, all the more so because the relevance of the dividing line between licit and illicit drugs upon which such policies have been founded is challenged by scientific data and practices.[19] Between repression and total freedom, the CCNE has proposed a new policy based on the scientifically sound distinction between *occasional use* and *abuse* resulting from compulsive consumption. The objective is that the population as a whole should be protected against the risks of developing drug addiction by legislation that on the one hand makes it possible to control products and access to them, in the best interests of public health, and on the other hand proportionately sanctions abuse and wrong done to others, while fully respecting individual liberty and dignity. *Use* (and nothing more) of a dangerous product should only be repressed if it takes place in public or when proof is produced that such action is harmful to others, particularly one's family.

Preventive policies must address the social and economic causes that aggravate the risk of drug addiction. Prevention must also be based on education for empowerment. Risks inherent to each substance must be neither exaggerated nor minimized. Prevention must also include structuring and regulating the use of substances affecting the central nervous system for the simple reason that in varying degrees they are all potentially dangerous. Each product should be ranked in terms of toxicity, therapeutic utility, risks of dependence due to consumption, danger of desocialization, and risks to third parties, and these should be reviewed periodically. Harm to oneself calls for medical rather than penal responses; harm done to others using certain products in public, incitement to consume, and, of course, trafficking calls for sanctions. A ranking of penalties should accordingly be established with a view to the severity of offenses or the potential for harm done—starting with a warning and culminating in imprisonment and fines, with a summons, temporary withdrawal of the driver's license, community service, and so on in between.

This is by no means a revolutionary enterprise; yet even this kind of modification to criminal justice and public health legislation can probably only be achieved progressively. Certainly, it should regularly take into account new data contributed by research, which must be amplified and encouraged. Along these lines, Switzerland has recently undertaken a public consultation, by referendum, on the continuation of heroin-prescription therapies and substitution treatments (still on an experimental basis) for drug addicts. The results of this ballot, held September 28, 1997, are astonishingly clear. Swiss citizens rejected, by a 71 percent majority, a constitutional amendment that would have stopped these programs. This shows that public perception of drug abuse may, after democratic debate, shift from repression to prevention and therapy.

Among the several recommendations and guidelines adopted by the CCNE, emphasis was placed on the social reinsertion of drug abusers and, on different grounds, on the access to products—an important issue to consider without hypocrisy and with full awareness of the extreme complexity of the geopolitical and economic aspects of distribution of these substances. It is quite certain that the clandestine market is unhealthy and harmful for both nonusers and users because of the insecurity it engenders. The changes cannot be strictly national. The responsibility of states is at stake, as well as the ethics of their international policies.

ENDNOTES

[1] See Barbara Meyerhoff's contribution in Kathleen Berrin, ed., *Art of the Huichol Indians* (New York: The Fine Arts Museums of San Francisco/Harry N. Abrams, Inc., 1978).

[2] Robert MacCoun and Peter Reuter, "Interpreting Drug Cannabis Policy: Reasoning by Analogy in the Legalization Debate," *Science* 278 (1997): 47–52.

[3] Jean-Pierre Changeux, *Neuronal Man* (1995; reprint, Princeton, N.J.: Princeton University Press, 1997).

[4] Humphrey Peter Rang and M. Maureen Dale, *Pharmacology* (New York: Churchill Livingstone, 1995).

[5] Claude Bernard, *Leçons sur les effets des substances toxiques et médicamenteuses* (Paris: Bailliére, 1857).

[6]John Newport Langley, "On the Reaction of Cells and of Nerve-endings to Certain Poisons Chiefly as Regards the Reaction of Striated Muscle to Nicotine and to Curare," *Journal of Physiology* 33 (1905): 374–413.

[7]J. Monod, J. Wyman, and J.-P. Changeux, "On the Nature of Allosteric Transitions: A Plausible Model," *Journal of Molecular Biology* 12 (1965): 88–118.

[8]Jean-Pierre Changeux, "The Acetylcholine Receptor: An 'Allosteric' Membrane Protein," *Harvey Lectures* 75 (1981): 85–254.

[9]Jean-Pierre Changeux, "Chemical Signaling in the Brain," *Scientific American* (November 1993): 58–62.

[10]H. Dohlman, M. Caron, and R. Lefkowitz, "A Family of Receptors Coupled to Guanine Nucleotide Regulatory Proteins," *Biochemistry* 26 (1987): 2657–2664.

[11]T. Robbins and B. Everitt, "Neurobehavioral Mechanisms of Reward and Motivation," *Current Opinion in Neurobiology* 6 (1996): 228–236; R. Wise, "Neurobiology of Addiction," *Current Opinion in Neurobiology* 6 (1996): 243–251.

[12]James Olds, "Self-Stimulation of the Brain," *Science* 127 (1958): 315–324.

[13]George F. Koob and M. Le Moal, "Drug Abuse: Hedonic Homeostatic Dysregulation," *Science* 278 (1997): 52–63.

[14]Arvid Carlsson, "Antipsychotic Drugs and Catecholamine Synapses," *Journal of Psychiatric Research* 11 (1974): 37–64.

[15]M. Picciotto et al., "Acetylcholine Receptors Containing β2-Subunit are Involved in the Reinforcing Properties of Nicotine," *Nature* 391 (1998): 173–177.

[16]Jean-Pierre Changeux, "Functional Architecture and Dynamics of the Nicotinic Acetylcholine Receptor: An Allosteric Ligand-gated Ion Channel," *Fidia Research Foundation Award* 4 (1990): 21–168.

[17]Eric Nestler, "Under Siege: The Brain on Opiates," *Neuron* 16 (1996): 897–900.

[18]E. Merlo Pich et al., "Common Neural Substrates for the Addictive Properties of Nicotine and Cocaine," *Science* 275 (1997): 83–86.

[17]See <http://62.160.32.15//ccne> for report no. 43 (November 23, 1994).

The very idea of manipulating human behavior seems to stir up both fears and wishful fantasies. These interfere with a commonsense evaluation of the issues. This emotional reaction to the idea of behavior manipulation must be evaluated at three different levels: first, as a barrier to the collecting of facts and to sensible assessment of the social problem; second, as a factor in the social acceptance of the technology; and, third, as a manipulative tool for modifying human behavior in its own right. The first of these must be considered before any other discussion. An emotional reaction to the idea of behavior control seems to lead to a short circuiting of the process of evaluation. Most discussion of behavior control begins with the possibility of a new technology and then either jumps to the desirability of an application of such techniques in the immediate future or to a possible mechanism of control of the technique to prevent its abuse. In these short-circuited discussions the leap from the idea to plans for immediate social action omits a review of the factual issues that would seem to be necessary for a more deliberate evaluation. Very often there is a failure to distinguish between facts and predictions, between facts and values, and between values and proposals for social action. This smearing of the status of propositions limits the usefulness of many such discussions.

—Gardner C. Quarton

"Deliberate Efforts to Control Human Behavior
and Modify Personality,"
from *Dædalus* Summer 1967,
"Toward the Year 2000: Work in Progress

Alexander A. Borbély and Giulio Tononi

The Quest for the Essence of Sleep

INTRODUCTION

S LEEP IS A REVERSIBLE STATE of reduced consciousness during which the processing of sensory input is minimal, coordinated behavior is abolished, and cognitive activities (thinking, planning, reflection) are suspended. Sleep must be exceedingly important, judging from the time that is spent in this state (for example, a sixty-year-old person has spent at least twenty years asleep). It must be important also in view of its ubiquitous occurrence in vertebrates. No human or animal has been shown to be able to dispense entirely with sleep, although the amount of sleep needed varies considerably. The drive for sleep is impressively manifested in sleep-deprivation experiments, where "sleep pressure" becomes overwhelming and the maintenance of waking is virtually impossible. Subjects who stayed awake for days lapse immediately into sleep when no longer supervised, and sleep-deprived soldiers are said to have fallen asleep while marching. Like hunger or thirst, the drive for sleep appears to satisfy an elementary need. However, unlike eating and drinking, the purpose for sleep remains obscure. Sleep is the one major biological process whose raison d'être has not yet been specified. Unraveling its essence constitutes a major challenge for biological research in general and for neuroscience in particular.

Alexander A. Borbély is Professor of Pharmacology at the Institute of Pharmacology at the University of Zurich.

Giulio Tononi is Senior Fellow in Experimental and Theoretical Neurobiology at The Neurosciences Institute.

THE PHENOMENOLOGY OF SLEEP

According to the myths of ancient Greece both gentle sleep (Hypnos) and pitiless death (Thanatos) are sons of the goddess of night (Nyx). It is interesting that today the influence these sons have on us is determined by a common electrophysiological procedure: Typical changes in the pattern of the electroencephalogram (EEG) serve to discriminate sleep and waking, and a complete absence of the EEG is the unambiguous sign of brain death. Modern sleep research is intimately linked to the ability to record the EEG. From the classical studies in the 1920s and 1930s it is known that human brain waves increase in amplitude and decrease in frequency during the passage from waking to deep sleep. In 1937 Helen Blake and R. W. Gerard demonstrated that both EEG slow waves and the threshold to sensory stimuli rise to a maximum within the first hour after the onset of sleep and then gradually decline in the subsequent hours.[1] Slow waves have proven to be valuable markers of sleep intensity and, indirectly, of the need for sleep. Models of sleep regulation were based on this measure (see below). The focus on the EEG did not only promote sleep research but also delayed its progress. Rapid eye movement (REM) sleep, a sleep state that occupies 20 to 25 percent of adult sleep, was discovered only in 1953.[2] The reason for this late discovery was that the EEG signs of this sleep state are inconspicuous and resemble the superficial sleep stage 1 that typically occurs after the onset of sleep. However, unlike stage 1, REM sleep exhibits rapid eye movement, which occurs phasically under the closed eyelids, and a loss of the tonus of voluntary muscles. In addition, autonomic nervous activity shows increased variations during REM sleep, which are manifested by fluctuations in heart rate, blood pressure, and respiration, as well as by erections in men. The periodic occurrence of the latter phenomenon during sleep had been reported by P. Ohlmeyer and his colleagues in 1944, but their association with a separate sleep state was not recognized.[3] The traditional part of sleep, which was known long before the discovery of REM sleep, is presently referred to by the somewhat deprecatory designation "non-REM sleep." Other terms for non-REM and REM sleep are orthodox and paradoxical sleep or quiet and active sleep, respectively. The latter term refers to the fact that functions controlled by the motor and autonomic nervous system are inactive in non-REM sleep.

Transitions between "quiet" non-REM sleep and "active" REM sleep do not occur in a haphazard fashion but, rather, periodically. In the adult human the non-REM/REM sleep cycle exhibits a periodicity of 90–100 minutes. Although Nathaniel Kleitman's notion of the basic rest-activity cycle (BRAC) is no longer shared by sleep scientists, the cyclic alternation of the two basic sleep states is still one of the hallmarks of sleep.[4] It is now considered to be a sleep-dependent process that does not persist during waking.[5]

Adult humans typically sleep during the night hours and are awake during the day. This is an adaptation to the time structure of society, where work, school, and most other activities are scheduled in the daytime hours, but it is also due to biological factors. Many functions of the organism are modulated by the circadian pacemaker, which generates a rhythm with a period close to twenty-four hours.[6] This endogenous rhythm is synchronized by the external twenty-four-hour cycle, and by light in particular. Core body temperature is one of the physiological variables undergoing a distinct circadian modulation, the maximum being situated in the late afternoon and the minimum in the early morning hours. Some hormones are secreted with a twenty-four-hour (circadian) pattern and therefore may serve as markers of a circadian phase. Thus the pineal hormone melatonin starts to be secreted during the late evening hours, reaching a maximum level during the night and declining in the morning to its low daytime level.[7] It is known that sleep propensity undergoes a marked circadian variation, which is roughly the inverse of the core body-temperature rhythm. The maximum of the circadian sleep propensity occurs in close proximity to the minimum of the core body temperature. Also, the REM sleep/non-REM sleep ratio is subject to a circadian modulation. When sleep propensity is at its maximum, the REM-sleep fraction of the 90–100 minute cycle is the largest.[8] Since under habitual conditions this maximum is situated in the early morning hours, REM sleep exhibits a rising trend during the nighttime sleep episode. As will be mentioned later, this trend is not only the result of the circadian rhythm but also of a progressive release from the inhibition by non-REM sleep.

Taken together, our former view of sleep as a singular state has evolved to the notion of a complex dynamic process (figure 18, plate 16) dominated by the cyclic alternation of two basic substates. One, non-REM sleep, exhibits an intensity dimension that is reflected by the preponderance of EEG slow waves. Hence, slow-wave activity can serve as a marker of the changes of non-REM sleep intensity throughout the sleep episode.[9] Its time course typically shows an initial maximum followed by a progressive decline. REM sleep, an activated sleep state with distinct phasic events, exerts marked influences on the autonomic nervous system as well as on the voluntary motor system. A further typical feature of human sleep is its preferential occurrence during a particular phase of the circadian cycle. Can the phenomenology of sleep be transcended to gain insight into its functional aspects? A promising avenue is the exploration of regulatory features.

SLEEP REGULATION

Sleep is a history-dependent regulated process. Prolonging wakefulness enhances sleep propensity, whereas excess sleep has the reverse effect. The drive to maintain an optimal level of sleep has been referred to as "sleep homeostasis."[10] An important aspect of homeostatic sleep regulation is that a deficit is compensated for not only by a prolongation of sleep but also by its intensification. As mentioned earlier, EEG slow-wave activity may serve as an electrophysiological marker of sleep intensity. Its decline during the sleep episode can be described by an exponential function.[11] Conversely, the rising part of the process during waking follows a saturating exponential function. This has been shown by nap studies in which the content of slow-wave activity within daytime sleep episodes increased in proportion to prior waking.[12] The buildup in sleep propensity progresses beyond its usual level if sleep initiation is postponed.

The exponential decline of slow-wave activity during sleep episodes had been formulated from a global viewpoint that did not take into account the non-REM/REM sleep cycle. A more detailed analysis of single non-REM sleep episodes revealed that in each episode slow-wave activity exhibits a gradual buildup until a peak level is reached; before the onset of REM sleep a precipitous decline occurs. Thus, not only the mean level of slow-wave activity within an episode but also its buildup rate is regulated by a homeostatic process.[13] The steep buildup of the first non-REM sleep episode becomes increasingly shallow over the subsequent episodes as the "sleep drive" dissipates. Prolonged waking enhances the intra-episodic buildup rate.[14]

The close association between slow-wave activity and prior waking or sleep has created mathematical models.[15] The two-process model of sleep regulation accounted for the timing and structure of sleep in various protocols by the interaction of a homeostatic (Process S) and a circadian (Process C) component.[16] Simulations based on elaborated versions of the model predicted quite accurately the changes of slow-wave activity during various sleep schedules, and the simulated data were validated against empirical results.[17]

Sleep had looked like a rather simple process until the discovery of REM sleep revealed a new dimension of complexity. Similarly, simulating the sleep process while disregarding REM sleep entirely or introducing it in the model as an external, data-derived parameter proved to be relatively straightforward. Attempts were made to simulate the non-REM/REM sleep cycle itself. These approaches were based on theoretical considerations and neurophysiological hypotheses.[18] However, none of the models has so far succeeded in accounting for three major features: the cyclic alternation of non-REM and REM sleep, the interaction between the two substates, and REM sleep homeostasis.[19] Both the reciprocal interaction model and the limit cycle model postulate that the non-REM/REM sleep cycle arises from the interaction of distinct neuronal populations. However, this explanation appears to be incomplete in view of the pacemaker role of pontine neurons, which by themselves can initiate the REM sleep process.[20] The non-REM sleep process therefore could be viewed as being periodically interrupted by REM sleep. This interpretation is consistent with the time course of slow-wave activity in depressive patients in whom REM sleep had been largely abolished by antidepressant drugs.[21] However, the REM sleep trigger is not a simple pacemaker, because in humans the regular interval between REM sleep episodes is determined by sleep time (i.e., by non-REM sleep) and the "clock" appears to be stopped during experimental waking episodes.[22] To complicate matters further, the prevention of REM sleep leads to a rising number of attempts to initiate REM sleep, an indication of the rapidly increasing propensity for this state.[23] However, when REM sleep is finally allowed to occur, its rebound is more limited than would be expected from the deficit incurred.[24] This suggests that different mechanisms may underlie the onset and the maintenance of REM sleep.

Turning now to the circadian component of sleep regulation, it has become clear that its mechanisms are separate from the homeostatic component. In animals with lesions of the suprachiasmatic nucleus (SCN), the site of the principal circadian pacemaker, the circadian sleep-wake cycle, along with other circadian rhythms, is abolished. Nevertheless, the capacity to increase sleep duration and sleep intensity in response to sleep deprivation remains intact.[25] Forced desynchrony experiments in humans allowed the separate analysis of the circadian and the homeostatic component.[26] This analysis confirmed that slow-wave activity is largely determined by prior sleep and waking, and that it is little influenced by the circadian pacemaker. Other sleep variables such as sleep continuity are determined by both components.

Despite the complexity of the different regulatory processes of sleep, one dominant theme emerges: the relationship between EEG slow-wave activity and the duration of prior sleep and waking. This relationship prevailed in all mammalian species hitherto examined; it applies to the 75 to 80 percent of human sleep time that is made up of non-REM sleep and proved to be largely impervious to circadian modulation as well as to the ultradian modulation by the periodically occurring REM sleep. What, then, is its physiological basis? Before seeking an answer in the cellular electrophysiology of sleep, we will examine variations of sleep duration and structure both among individuals and among species in the hope of gaining further insights into relevant basic processes.

INDIVIDUAL VARIATION IN SLEEP DURATION

There are famous short sleepers and long sleepers. Edison is reported to have slept only four to six hours per night, whereas Einstein enjoyed spending up to ten hours in bed. An extreme case of a short sleeper was reported by the English sleep researcher Ray Meddis and his colleagues: a seventy-year-old retired nurse claimed to get along on only one hour of sleep a night.[27] She did not feel tired, and spent the night writing and painting. Her claim was confirmed by polygraphic recordings in the sleep laboratory. Deep sleep (i.e., slow-wave sleep) made up almost half of her sleep time. This is an unusually high proportion for a seventy-year-old person and attests to the high priority of slow-wave sleep. Experimental sleep curtailment in other studies is consistent with this conclusion. Recently, healthy young short sleepers were compared to long sleepers to investigate the sleep-regulating mechanisms in the two groups.[28] The principal question was whether the homeostatic process underlying non-REM sleep regulation had the same time constant in both types of sleepers. Based on theoretical considerations derived from the two-process model it was predicted that short sleepers would respond to sleep deprivation by a smaller increase in slow-wave activity than long sleepers. This prediction was confirmed by the experiment (figure 19, plate 16). A further result was that short sleepers tended to be less affected by sleep deprivation than long sleepers. The amount of slow-wave sleep was comparable in the two groups, despite the fact that the total sleep time in the long sleepers was 50 percent higher. The results confirm the high priority of slow-wave sleep and demonstrate that short sleepers appear to live at a higher level of "sleep pressure" than long sleepers. They seem to tolerate sleep restriction better than long sleepers. Why this is the case is still an open question; we obviously need a better understanding of the processes linking sleep need to the underlying physiological processes. The issue of sleep loss is not only of theoretical interest. It has been estimated that errors due to sleep deprivation and sleepiness cause 25,000 deaths and 2.5 million disabling injuries and cost over $56 billion each year in the United States alone.[29]

THE COMPARATIVE PHYSIOLOGY APPROACH

The different mammals that have been investigated so far all show the typical features of sleep: an increase of slow-wave activity in response to sleep loss, the cyclic alternation of non-REM sleep and REM sleep, and the circadian modulation of the sleep process.[30] The latter is highly variable between species. In some animals (i.e., guinea pigs) a circadian rhythm is barely detectable.[31] Unique conditions prevail in those rodents that lower their metabolism and lapse into a prolonged sleeplike state. Hibernating species, such as the ground squirrel, reduce their body temperature to 4°C for days. Their hibernation period is periodically interrupted by brief euthermic episodes. The Djungarian hamster exhibits a milder form of hypometabolism. It passes into a daily torpor for several hours, where its body temperature declines to that of room temperature. Since natural sleep is also accompanied by a slight reduction of metabolism and body temperature, it was interesting to compare it to the more prominent hypometabolic states. An unexpected observation was that sleep occurring immediately after the emergence from hibernation or torpor showed an increase in slow-wave activity, the typical response to prolonged waking.[32] Moreover, the enhancement of slow-wave activity was proportional to the duration of the episode spent in hibernation or daily torpor. This indicates not only that hypometabolic states are functionally dissimilar from natural sleep but also that they appear to give rise to a sleep debt. The reduction of metabolism does not arrest the process leading to a rise in slow-wave propensity.

When did sleep make its first appearance during evolution? This question is not easily addressed, because the EEG criteria of mammalian sleep cannot be applied to animals with simpler nervous systems. Sleeplike behavioral states are widely known to exist in different classes of vertebrates as well as in some invertebrates (e.g., the moth, bee, and crayfish).[33] Is there evidence for a history-dependent regulation of a sleeplike state that is comparable to mammalian sleep homeostasis? To examine this question, two invertebrate species, the scorpion and the cockroach, were "rest-deprived" by keeping the animals active for several hours.[34] Both species exhibited an increase in rest episodes after these manipulations. The results are consistent with an early appearance of homeostatic regulatory mechanisms, which compensate excessive activity by a prolonged sleep-like resting state. One of the interesting future questions to be investigated is whether changes at the molecular level associated with sleep (see below) can be found during sleeplike states of simple organisms.

GLOBAL VERSUS LOCAL SLEEP

Sleep in mammals is defined principally on the basis of neurophysiological criteria, with the EEG serving as the principal state marker. Thus, it is the specific mode of brain activity that is most intimately linked to the sleep process. Does this process always encompass the brain as a whole, or is it conceivable that it is present to a variable extent in different brain regions? If so, this may help to shed light on the events that ultimately give rise to sleep. So far we know that the waking state leads to a buildup of slow-wave propensity and that this buildup can be described by a saturating exponential function (figure 19). However, it is not known if a specific aspect of waking is the crucial factor. Motor activity does not seem to be indispensable. In the rat, forced locomotion or voluntary wheel-running during waking hours had similar effects on slow-wave activity, as did sleep deprivation achieved by gentle handling.[35] In humans, staying awake and virtually immobile in a semisupine position—the typical situation of "constant routine" protocols—showed effects on slow-wave activity similar to an upright and mobile waking period.[36] Studies in subjects undergoing large physical exertion (e.g., marathon running) have failed to document unambiguously the effects on slow-wave sleep.[37] Finally, as has been mentioned before, animals spending time in sleeplike hypometabolic states of hibernation or daily torpor accumulate a sleep need as reflected by enhanced slow-wave propensity. Taken together, these observations suggest that sleep propensity increases during any state other than sleep. Accordingly, in a recent version of the two-process model, it was posited that there is a continuous rise in the sleep promoting Process S and that an antagonistic process is activated only in the presence of slow waves.[38] This hypothesis is consistent with the finding that the suppression of slow waves in the first part of sleep by acoustic stimuli leads to a rebound of slow waves in the second, undisturbed part of sleep.[39]

The view that the sleep process encompasses necessarily the entire brain is contradicted by results obtained in aquatic mammals. The dolphin, for example, exhibits deep non-REM sleep with dominant slow waves only in one hemisphere at a time, whereas light non-REM sleep can be present in both hemispheres simultaneously.[40] Moreover, when the occurrence of deep non-REM sleep was prevented in one hemisphere, a rebound of slow waves was observed only in the deprived hemisphere.[41] These results demonstrate that not only the occurrence but also the regulation of slow waves may be restricted to a specific part of the brain.

There is recent evidence from studies in monkeys that the process of falling asleep may not occur synchronously in the entire brain.[42] The animals were trained to perform a visual search task while the activity of their cortical neurons was being recorded. At times the monkeys became drowsy, and the neurons in the extrastriate cortical area V4 exhibited the typical sleeplike burst-pause pattern and no longer responded to the visual stimuli. In spite of this "neuronal sleep," the animals continued to perform the visual task, which indicates that their primary visual cortex was still responsive. Thus, a portion of the cerebral cortex may "fall asleep" while another is still performing a behavioral discrimination that has become automatic.

The possibility that humans may also show regionally modulated sleep signs was recently examined in the following experiment. Subjects were administered prolonged vibratory stimuli to one hand during the waking hours preceding sleep in order to activate the contralateral somatosensory cortex.[43] The question was whether daytime activation of a specific brain region has regional repercussions on the EEG during subsequent sleep. This was found to be the case after stimulation of the dominant (right) hand. The EEG recorded from the central lead overlying the somatosensory cortex exhibited a shift in power towards the contralateral side during the first hour of sleep. Such changes were not present at frontal, parietal, and occipital derivations. In the central derivation, the shift in power occurred in the low-frequency range of the spectrum. It must be added that the effect was small and restricted to right-hand stimulation. Nevertheless, it demonstrated for the first time that human sleep has a local, use-dependent facet—an aspect of sleep that had been postulated by J. M. Krueger and F. Obál.[44]

A regional facet of sleep could be expected to emerge also during normal sleep, as not all cortical areas are involved to the same extent in the waking process. Recent experiments have shown that a sleep deficit impairs primarily high-level cognitive skills, which depend on frontal lobe function.[45] Patients with lesions of the prefrontal cortex suffer from deficits that include distraction by irrelevant stimuli, diminished word fluency, flat intonation of speech, impaired divergent thinking, apathy, and childish humor.[46] Subjects foregoing sleep exhibit similar symptoms. Therefore, it may be more than a coincidence that the prevalence of slow waves is maximal at frontal EEG derivations in the initial part of sleep.[47] The frontal predominance in slow waves was present in the first and second non-REM sleep episodes and vanished in the subsequent parts of sleep. This finding may indicate that a putative recuperative process of sleep, of which slow waves are a marker, occurs at a higher rate in frontal areas. Conversely, the brain functions mediated by these areas are the first to be jeopardized by the absence of sleep.[48] In summary, EEG slow waves are not only markers of global non-REM sleep-intensity, which is closely related to the duration of prior sleep and waking, but also exhibit a regional specificity. The "slow-wave mode" of sleep is restricted to one hemisphere in the dolphin, and in humans it may be enhanced by a selective (i.e., vibration experiments) or natural, preferential activation (i.e., frontal cortex) of specific brain regions. Based on this electrophysiological sleep sign, it appears that sleep is not only a global phenomenon but also has local, use-dependent facets.

CELLULAR CORRELATES OF SLEEP

How small a portion of the brain is capable of sleep? If we could answer this question, we would likely understand a fundamental principle about the nature of sleep. More than thirty years ago

Giuseppe Moruzzi, one of the pioneers of the neurophysiology of sleep, argued that when cells or synapses that support conscious activity undergo plastic changes during waking, they slowly accumulate some chemical change. He suggested that it is because of such cellular or synaptic changes that we need to sleep, although he did not speculate what such changes might consist of.[49] He recognized, of course, that in the normal course of events cells and synapses do not sleep at random. Centralized mechanisms, located in the brain stem and hypothalamus, ensure that our brain falls asleep more or less at the same time; otherwise, "we would spend most of our life in a dormiveglia."

Long after Moruzzi's prophetic remarks, finding key cellular correlates of sleep and waking is becoming one of the main goals of present-day neurobiological investigations of sleep. The identification of such correlates would provide a rationale for investigating specific aspects of sleep and waking in simplified preparations, such as brain slices or cell cultures. Also, a robust cellular correlate of sleep would lend itself to comparisons across ontogeny and phylogeny, as well as in dissociated or pathological states. Although searching for cellular correlates of integrative functions may look excessively reductionistic, there are successful precedents for this approach. For example, it has recently been shown that a single cell of the suprachiasmatic nucleus is a minimal brain unit capable of exhibiting circadian rhythmicity.[50]

THE CELLULAR ELECTROPHYSIOLOGY OF SLEEP

Most of our knowledge about the cellular basis of sleep over the last forty years has been gained from recording the electrical activity of single neurons. The very first studies showed that, contrary to expectations, single neurons in the cerebral cortex do not stop firing during sleep.[51] The spontaneous rate of firing is only slightly lower in non-REM sleep than in waking, and it is often higher in REM sleep. For most cortical neurons, the simpleminded hypothesis that sleep provides rest by mere inactivity, as relaxation does for muscle cells, is untenable. Several aspects of neural activity, however, differ markedly between waking and sleep. These differences have been studied in recent years by examining the thalamocortical system.

Sensory information impinging upon the cerebral cortex is funneled through the thalamus. During waking, neurons in the brain stem reticular formation exert an excitatory influence on neurons in the thalamus. Consequently, the membrane potential is kept in a state of relative depolarization, and activated nerve cells discharge tonically in single spikes. After sleep onset, the activating influence of brain stem reticular neurons diminishes and as a result the membrane potential of thalamic and cortical neurons becomes more negative. Due to this hyperpolarization, both thalamic and cortical neurons generate repetitive burst discharges followed by long pauses during which their membrane potential is very negative.[52] The ability of neurons to respond to extrinsic stimuli is reduced. This results in the disconnection of the brain's intrinsic activity from outside events. How do these cellular changes in the brain relate to the EEG pattern that is recorded from the scalp? A fascinating new result is that the membrane potential of thalamocortical neurons and the cortical EEG exhibit similar fluctuations. As sleep deepens the nerve cells in the thalamus become increasingly hyperpolarized, with fluctuations in their membrane potentials occurring first in the frequency range of the sleep spindles and then in the range of slow waves. The same typical sequence of changes is seen in the EEG as non-REM sleep progresses from a superficial stage that is characterized by the periodic occurrence of sleep spindles to deep sleep, where slow waves predominate. Recently, an even slower rhythm with a frequency below 1 Hz was described at the cellular level and then discovered also in the EEG.[53] These new developments are important for sleep research because they make it possible to identify the neurophysiological mechanisms underlying the sleep EEG.

Not all neurons exhibit the typical low firing rate during non-REM sleep and a high firing rate during REM sleep. A few cell groups in the brain stem reduce their activity in non-REM sleep and become almost silent in REM sleep. These particular neurons send diffuse projections to the entire brain, influencing their targets by releasing the neuromodulators norepinephrine, serotonin, or histamine. For example, the noradrenergic neurons in the locus coeruleus and the serotonergic neurons in the raphe system fire tonically during waking, discharge at much lower levels during non-REM sleep, and cease firing during REM sleep. These neurons typically discharge in a transient, phasic manner whenever the animal encounters a salient event. This modulation is lost during sleep. The distinctive firing pattern of these neurons indicates that they may be closely associated with the behavioral state of the animal.

MOLECULAR ASPECTS OF SLEEP

Compared to our understanding of electrophysiological correlates of sleep and waking, the search for correlates at the molecular level is still in its infancy. Perhaps the most striking finding in recent years has been that the transition from sleep to waking is accompanied by the activation of the expression of immediate-early genes in many brain regions. These are special genes that can act as transcription factors to regulate the expression of other genes; they are induced by various electrical or chemical signals to which neural cells are exposed. It was found that, after a few hours of sleep, the expression of these transcription factors in a rat's brain is very low or absent. However, if a rat was either spontaneously awake or sleep deprived for a few hours, the expression of c-Fos, NGFI-A, and other immediate-early genes was high in the cortex, hippocampus, and other brain regions.[54] The increased expression of immediate-early genes is accompanied by an increased phosphorylation of constitutive transcription factors, such as CREB, that have been involved in learning and memory. It is also accompanied by a dramatically increased level of phosphorylation of other, as yet unidentified, cellular proteins.[55]

Cells expressing c-Fos or other immediate-early gene products after periods of waking are distributed over the entire cerebral cortex. Nevertheless, they represent only a subset of all neurons in any given area. At present, it is not clear which cells express immediate-early genes and why. It is possible that some neurons are simply in a better position than others for a strong response to the complex combination of electrical and chemical signals that they are exposed to during waking. A stronger activation of intracellular signaling pathways in these selected neurons will lead to the activation of their transcriptional machinery through the action of immediate-early genes. As indicated by studies in organisms as diverse as the sea slug (*Aplysia*), the fruit fly (*Drosophila*), and the mouse, the activation of transcription is a necessary step for the transition from short-term to long-term changes in neuronal and synaptic function and therefore for the acquisition of new memories.[56]

THE ROLE OF DIFFUSE ASCENDING SYSTEMS

The striking difference between sleep and waking in the expression of immediate-early genes and in the phosphorylation of certain proteins has prompted the search for the underlying neural mechanisms. Recent studies have shown that a critical parameter may be the activity of neuromodulatory systems with diffuse projections whose neurons discharge during waking and not during sleep.[57] When salient events during waking cause a burst of activity in the locus coeruleus, the diffuse release of norepinephrine appears to act through a complicated biochemical pathway that leads to the phosphorylation of transcription factors and to the induction of Fos and other immediate-early genes in many brain areas. During sleep, when the cells in the locus coeruleus discharge at a low level or become silent, this pathway becomes ineffective.

The silence of diffuse ascending systems may explain why we cannot acquire new information during sleep. Despite claims to the contrary in the popular press, there is very little support for the notion that we may learn anything during sleep, whether in non-REM or REM sleep. The reason may be that the inactivity of these systems during sleep prevents neurons from undergoing plastic changes. This is consistent with evidence from several different fields suggesting that a concurrent activation of neuromodulatory systems is required for triggering long-term changes of neural function.[58] It also makes functional sense, because the brain can learn how to successfully adapt to the environment only if it is awake.

The reduced neuronal firing during sleep in diffuse neuromodulatory systems that release norepinephrine, serotonin, and histamine has been demonstrated in rats, cats, and monkeys.[59] While the effects on target cells may be quite complicated, a general pattern emerges. For example, there is evidence that norepinephrine promotes the responsiveness of neurons to extrinsic stimuli, enables plastic changes both during development and in the adult stage, and increases the rate of metabolic processes.[60] On a time scale of fractions of seconds, norepinephrine release enhances the readiness of neurons to respond to external, salient signals by increasing their "signal-to-noise" ratio. As we have seen, norepinephrine may also permit the animal to respond to salient extrinsic signals with long-term changes in neural function through its action on protein phosphorylation and on the activation of transcription factors. Finally, by triggering cellular metabolic changes, norepinephrine may increase the ability of neurons to face the increased energetic demands required by the interaction with the environment. If a consistent picture were to emerge for other monoamines, such as serotonin and histamine, a low release of these neuromodulators may turn out to be an essential cellular signature of sleep. As a signature, it would have the important advantage of being common to both non-REM and REM sleep. Furthermore, it would connect a basic behavioral feature of sleep, the reduced responsiveness of the animal to external stimuli, to a cellular feature that accompanies the reduction of monoaminergic activity, the reduced cellular responsiveness to external stimuli in terms of both activity and plasticity.

SLEEP: A TIME FOR RESTORATION OR A TIME FOR STIMULATION?

If sleep is a time in which not only the entire organism but individual nerve cells become less responsive to the outside world, it is natural to ask what purpose such isolation might serve. What sleep is for, however, is the most embarrassing question one could ask a sleep researcher. Many theories have been formulated; none has gained widespread support. In small mammals, sleep may save precious energy when the prospects for foraging are not good, but in other species this explanation is not satisfactory. In despair, some have suggested that sleep may simply have kept our ancestors out of trouble at night. But it is unclear why, if this were the case, evolution would have gone to such length in developing complicated adaptations such as unihemispheric sleep in the dolphin. There are elegant theories that assign complementary roles to non-REM and REM sleep, yet other equally elegant theories maintain that non-REM and REM sleep serve a similar purpose. The question is, what purpose? In the last part of this essay, we will move from the secure terrain of experiments to the uncertain ground of hypotheses.

Artist, poets, and common folk alike have always assumed that sleep is rest for both body and brain—a welcome period of recovery, whether from physical or mental exertion. Yet, a simple observation suggests that sleep is, at its essence, for the brain. If we lie awake but immobile for an entire night, in the morning our muscles are relaxed but our mind is not, and our sense of well-being is lost. While quiet waking may be an equivalent to sleep for most bodily organs, it is not so for the brain.

That sleep is for the brain and not for the body was argued convincingly by Moruzzi; sleep is not for the muscles, not for the visceral organs, not even for the autonomic nervous system or the spinal cord. In fact, he suggested that "sleep concerns primarily not the whole cerebrum, nor even the entire neocortex, but only those neurons or synapses, and possibly glia cells, which during wakefulness are responsible for, or related to, the brain functions concerned with conscious behavior."[61] As we have seen, this view is partially supported by the observation that the most notable dysfunctions observed after sleep deprivation have to do with the prefrontal cortex. On these premises, Moruzzi then presented what still counts as one of the most interesting restorative hypotheses of sleep: "Sleep recovery would be responsible for the stability of the physicochemical properties of those brain structures that are affected by, or contribute to, the plastic changes occurring during the waking state."[62] A waking brain is conscious. Sleep, or perhaps certain phases of sleep, may offer the blessing of unconsciousness or at least the liberation from the tyranny of memory.

CELLULAR AND SUBCELLULAR RESTORATION

The idea that sleep restores the brain to a condition that was present at the beginning of the previous period of waking has led to the formulation of scores of so-called restorative theories of sleep. Some theories have considered what brain constituents may be depleted during waking that need to be replenished during sleep; others have proposed that some harmful compound may be accumulating that needs to be eliminated. The differences among these theories lie in what needs to be restored or eliminated, how this occurs, and why.

While it is not possible to review all such theories, it is worth considering whether taking an unabashedly cellular approach may offer some fresh insights and, most importantly, suggest new experimental tests. The availability of such powerful methods as nuclear magnetic resonance and of sophisticated biochemical assays should soon permit the reliable evaluation of changes in the concentration of specific cellular components. Recent developments in molecular biology offer the unique opportunity to study cellular regulatory processes in a systematic way, as long as they are reflected in changes in gene expression inside the cell. Techniques such as subtractive hybridization and differential display can be employed to determine, at least in principle, all the genes that modify their expression between sleep and waking.[63] Identifying these genes should point to cellular homeostatic processes that may have been activated by sleep and waking, thus paving the way for direct experimental tests.

A recent example of a hypothesis poised at the cellular level is the idea that sleep may be associated with the restoration of brain glycogen.[64] In the muscle and liver, glycogen represents an important energy store. In the brain, glycogen is contained exclusively in glial cells called astrocytes. During waking, the release of the neuromodulators norepinephrine, serotonin, and histamine produces glycogenolysis, which may be a way to provide neurons with increased fuel at times of increased metabolic demand. Increased metabolic demand may cause a transient deficit in cellular energy charge, which translates to a higher concentration of adenosine. According to the hypothesis, adenosine would help promote and increase slow waves in non-REM sleep. Due to the reduced metabolic demand and the reduced release of norepinephrine, serotonin, and histamine, sleep would be a propitious time for the replenishment of the glycogen store via glycogen-synthesis. Whether or not this hypothesis will hold true, it has the beauty of being both simple and general, a welcome property considering the universality of sleep.

Another possibility is related to Moruzzi's idea that sleep may serve to restore the function of certain synapses in the cerebral cortex. Each cortical axon makes thousands of synaptic contacts with target cells located in either the same cortical area or in different areas. The synapse between neurons is the critical site where signal transmission occurs and where plastic modifications may take place, which are the substrate for learning and memory. To perform their tasks, synaptic terminals contain two main constituents: synaptic vesicles and mitochondria. Synaptic vesicles are full of the neurotransmitter glutamate, which after its release into the synaptic cleft excites postsynaptic target cells. Mitochondria are ubiquitous organelles that, through the respiratory chain, provide the necessary energy to keep synaptic transmission going.

During quiet waking, metabolic rates in the brain are higher than during non-REM sleep by up to 30 percent.[65] The higher energy consumption is probably due to the increased work load of ion pumps that counteract the increased neuronal depolarization of waking. We do not know whether glutamate contained in synaptic vesicles is released at higher levels during waking than during sleep.[66] However, the depolarization observed during waking suggests a net increase of excitatory neurotransmission during waking and a decrease during sleep.[67] At any rate, it would appear that during waking the mitochondria, the energy furnaces of neurons, must work at a higher speed. The results of a recent systematic survey of gene expression in the brain of sleeping and waking animals using differential display provide some pertinent observations. It was found that waking is associated with levels of mitochondrial mRNA that are higher than in sleep by 20 to 30 percent.[68] This is a large change considering that these mRNAs are expressed ubiquitously and at high levels. Mitochondrial mRNAs are transcribed and translated within the mitochondrion into subunits of key respiratory enzymes, such as cytochrome oxidase, cytochrome B, and NADH dehydrogenase. It is possible that this increase may represent a local regulatory response of mitochondria. Such an increase may transiently help the synaptic terminals face the increased energy requirements of waking. In the long run, however, mitochondria need crucial constituents that can only be synthesized in the cell body and transported via axonal transport.

Issues related to the efficiency of synaptic transmission are of particular interest because higher brains face a unique logistic problem. Billions and billions of synaptic terminals are far away from the cell body, connected to it by an axon that is very thin and very long. How can it be guaranteed that mitochondria are kept functioning efficiently or that synaptic vesicles are constantly replenished at such distance from the cell body? Is it the case that mitochondria are less challenged during sleep, or that neurotransmitter release is reduced, or that transport of critical mitochondrial and vesicle constituents along the axon is favored? Indeed, one of the most robust correlations in the phylogeny of sleep is that the length of the non-REM/REM sleep cycle is closely related to brain weight or size.[69] Unfortunately, no experiment as yet has addressed the possibility that non-REM and REM sleep may be important for cellular transport processes. We do not even know whether key constituents of synaptic terminals, such as mitochondria and synaptic vesicles containing glutamate, are seriously at risk of exhaustion under conditions of sleep deprivation. If that were the case, one would expect that prolonged waking may result in synaptic fatigue and eventually synaptic failure. Remarkably, although we all associate weariness and sluggishness of thought with prolonged sleep deprivation, no study has addressed this possibility in vivo.

STIMULATION AND CONSOLIDATION

In our present state of ignorance, it is customary to encounter hypotheses about the functions of sleep that are diametrically opposed to each other. For example, it has been suggested by some that REM sleep may help consolidate memories and by others that it may serve to erase them.[70] Similarly, alongside the hypotheses that consider sleep as a time for restoration, there are those that view sleep as a time for stimulation. The latter hypotheses are particularly appealing with respect to the developmental aspects of sleep. The fetus spends most of its time in states of active sleep that resemble REM sleep. Activity-dependent processes are essential for axonal growth, synaptic remodeling, and the refinement of the connectivity. In the absence of extrinsic stimulation, the intrinsically generated activity of sleep, which is often oscillatory in nature, might promote the growth and maturation of neural circuits.[71]

In the adult, waking activity provides plenty of extrinsic stimulation. Nevertheless, the intrinsic patterns of the activity of sleep may still serve a similar purpose by complementing, rather than substituting for, the stimulation of waking. For instance, it has been hypothesized that REM sleep may serve to maintain the fitness of those neural circuits that, although phylogenetically important, are infrequently exercised during the life of an individual.[72] Others have suggested that both non-REM and REM sleep may serve to stimulate or stabilize synapses that are insufficiently used during waking and that might otherwise atrophy and die.[73]

Another possibility is that sleep may be necessary for the consolidation of selective events that occurred during waking. We have seen that, in a subset of neurons widely distributed over the cerebral cortex, waking leads to the activation of the transcriptional machinery, as indicated by the phosphorylation of certain transcription factors and by the expression of immediate-early genes. The activation of transcription may lead to long-term changes in synaptic strength that involve extensive synaptic remodeling, including the formation of new synaptic contacts through axonal sprouting and dendritic branching.[74] These cellular events are likely to require the increased synthesis of various constituents necessary for building and maintaining synapses, including synaptic vesicles and mitochondria. Furthermore, newly formed synapses require some form of neural activity in order to be stabilized.[75] It is natural to wonder, then, whether non-REM sleep might offer a biochemical or electrical environment that promotes transport processes, stabilizes synaptic remodeling, or favors growth in those neurons that have been selected for transcription during waking. In this view, REM sleep could provide newly formed synapses with a pattern of stimulation closer to that found during waking hours. Since neuromodulatory systems with diffuse projections are silent, however, such stimulation would occur without simultaneously triggering neuronal selection at a time when the organism is not adapting to its environment. In sequence, the two sleep states may promote the consolidation of those circuits that have been selected while awake.

A recent study examined the activity, in both waking and sleep, of hippocampal neurons that fire when a rat occupies a certain location in space.[76] Cells that fired together when the animal occupied particular spatial locations showed an increased tendency to fire together during subsequent non-REM sleep, in comparison to sleep episodes preceding the behavioral tasks. This effect declined gradually during each postbehavior sleep session. These findings have been interpreted to indicate that sleep may reactivate patterns of activity experienced during waking. Such reactivation during sleep may help to consolidate the synaptic changes triggered by waking behavior.

At the present stage, hypotheses such as these are as intriguing as they are premature. It might as well be proposed that rather than consolidating the changes triggered by previous waking, brain activation during sleep could promote random remodeling of brain circuitry to provide a fresh substrate for future selective processes. The absence of conclusive data still allows for the possibility that sleep exists for one thing as much as for its opposite—for restoration as much as for stimulation, for consolidation as much as for variation. Although our ignorance about the purpose of such an important part of our life despite many decades of research may seem disappointing, it is perhaps not surprising. Sleep involves consciousness and behavior, brain and body, psychology and physiology. It is a paradigmatically integrative phenomenon that is bound to be highly complex and therefore challenging to study. It would appear, however, that its experimental analysis in cellular and molecular terms and the further exploration of its regulatory features may be a promising way to proceed.

ACKNOWLEDGMENTS

The research of Alexander A. Borbély referred to in this paper is supported by the Swiss National Science Foundation and the Human Frontiers Science Program. The research of Giulio Tononi referred to in this paper was supported by

The Neurosciences Research Foundation (1995–present) and by grants of the Ministero dell' Universita' e della Ricerca Scientifica e Tecnologica and the Agenzia Spaziale Italiana (1992–1995). The authors thank Ms. Yvonne Maeder for her help with this manuscript.

ENDNOTES

[1] H. Blake and R. W. Gerard, "Brain Potentials during Sleep," *American Journal of Physiology* 119 (1937): 692–703.

[2] E. Aserinsky and N. Kleitman, "Regularly Occurring Periods of Eye Motility, and Concomitant Phenomena, during Sleep," *Science* 118 (1953): 273–274.

[3] P. Ohlmeyer and H. Huellstrung, "Periodische Vorgänge im Schlaf," *Pflügers Archiv (European Journal of Physiology)* 248 (1944): 559–560.

[4] N. Kleitman, *Sleep and Wakefulness* (Chicago, Ill.: The University of Chicago Press, 1963).

[5] See V. Brezinová, "Sleep Cycle Content and Sleep Cycle Duration," *Electroencephalography and Clinical Neurophysiology* 36 (1974): 275–282; H. Schulz and W. Tetzlaff, "Distribution of REM Latencies after Sleep Interruption in Depressive Patients and Control Subjects," *Biological Psychiatry* 17 (1982): 1367–1376; K. Fukuda, A. Miyashita, and M. Inugami, "Sleep Onset REM Periods Observed after Sleep Interruption in Normal Short and Normal Long Sleeping Subjects," *Electroencephalography and Clinical Neurophysiology* 67 (1987): 508–513; A. Miyashita, K. Fukuda, and M. Inugami, "Effects of Sleep Interruption on REM-NREM Cycle in Nocturnal Human Sleep," *Electroencephalography and Clinical Neurophysiology* 73 (1989): 107–116.

[6] For early overviews of human studies see J. Aschoff, "Circadian Rhythms in Man," *Science* 148 (1965): 1427–1432; and R. A. Wever, *The Circadian System of Man: Results of Experiments under Temporal Isolation* (New York: Springer-Verlag, 1979). For recent reviews see C. A. Czeisler, "The Effect of Light on the Human Circadian Pacemaker," *Ciba Foundation Symposium* 183 (1995): 254–290, discussion 290–302; and R. Y. Moore, "Circadian Rhythms: Basic Neurobiology and Clinical Applications," *Annual Review of Medicine* 48 (1997): 253–266.

[7] A. Brzezinski, "Melatonin in Humans," *New England Journal of Medicine* 336 (1997): 186–195.

[8] C. A. Czeisler et al., "Timing of REM Sleep is Coupled to the Circadian Rhythm of Body Temperature in Man," *Sleep* 2 (1980): 329–346; D.-J. Dijk and C. A. Czeisler, "Contribution of the Circadian Pacemaker and the Sleep Homeostat to Sleep Propensity, Sleep Structure and Electroencephalographic Slow-Waves and Sleep Spindle Activity in Humans," *Journal of Neuroscience* 15 (1995): 3526–3538.

[9] Slow-wave activity is represented by the spectral power density in the 0.75–4.5 Hz range; see B. A. Geering et al., "Period-Amplitude Analysis and Power

Spectral Analysis: A Comparison Based on All-Night Sleep EEG Recordings," *Journal of Sleep Research* 2 (1993): 121–129.

[10]Homeostasis has been defined as "the coordinated physiological processes which maintain most of the steady states in the organism." W. B. Cannon, *The Wisdom of the Body* (New York: Norton, 1939). The term "sleep homeostasis" refers to the sleep-wake dependent aspect of sleep regulation. A. A. Borbély, "Sleep: Circadian Rhythm Versus Recovery Process," in M. Koukkou et al., eds., *Functional States of the Brain: Their Determinants* (Amsterdam: Elsevier, 1980), 151–161. Thus homeostatic mechanisms counteract deviations of sleep from an average "reference level" by augmenting sleep propensity when sleep is curtailed or absent, and reducing sleep propensity in response to excess sleep.

[11]A. A. Borbély et al., "Sleep Deprivation: Effect on Sleep Stages and EEG Power Density in Man," *Electroencephalography and Clinical Neurophysiology* 51 (1981): 483–493.

[12]D.-J. Dijk, D. G. M. Beersma, and S. Daan, "EEG Power Density during Nap Sleep: Reflection of an Hourglass Measuring the Duration of Prior Wakefulness," *Journal of Biological Rhythms* 2 (1987): 207–219.

[13]P. Achermann and A. A. Borbély, "Simulation of Human Sleep: Ultradian Dynamics of EEG Slow-Wave Activity," *Journal of Biological Rhythms* 5 (1990): 141–157; P. Achermann et al., "A Model of Human Sleep Homeostasis Based on EEG Slow-Wave Activity: Quantitative Comparison of Data and Simulations," *Brain Research Bulletin* 31 (1993): 97–113.

[14]D.-J. Dijk, B. Hayes, and C. A. Czeisler, "Dynamics of Electroencephalographic Sleep Spindles and Slow Wave Activity in Men: Effect of Sleep Deprivation," *Brain Research* 626 (1993): 190–199.

[15]A. A. Borbély and P. Acherman, "Concepts and Models of Sleep Regulation: An Overview," *Journal of Sleep Research* 1 (1992): 63–79.

[16]A. A. Borbély, "A Two-Process Model of Sleep," *Human Neurobiology* 1 (1982): 195–204; and A. A. Borbély, "Sleep Homeostasis and Models of Sleep Regulation," in M. H. Kryger et al., eds., *Principles and Practice of Sleep Medicine*, 2d ed. (Philadelphia, Pa.: Saunders, 1994), 309–320; and S. Daan, D. G. M. Beersma, and A. A. Borbély, "Timing of Human Sleep: Recovery Process Gated by a Circadian Pacemaker," *American Journal of Physiology* 246 (1984): R161–R183.

[17]Achermann et al., "A Model of Human Sleep Homeostasis Based on EEG Slow-Wave Activity."

[18]R. W. McCarley and J. A. Hobson, "Neuronal Excitability Modulation over the Sleep Cycle: A Structural and Mathematical Model," *Science* 189 (1975): 58–60; H. Schulz et al., "The REM-NREM Sleep Cycle: Renewal Process or Periodically Driven Process?" *Sleep* 2 (1980): 319–328; R. W. McCarley and S. Massaquoi, "A Limit Cycle Mathematical Model of the REM Sleep Oscillator System," *American Journal of Physiology* 251 (1986): R1011–R1029; and S. Massaquoi and R. W. McCarley, "Extension of the Limit Cycle Reciprocal Interaction Model of REM Cycle Control: An Integrated Sleep Control Model," *Journal of Sleep Research* 1 (1992): 138–143.

[19]The term REM sleep homeostasis is used to denote the rebound of REM sleep after deprivation. However, in contrast to non-REM sleep there is no evidence that excessive REM sleep is compensated by a subsequent reduction.

[20]M. Jouvet, C. Buda, and J. P. Sastre, "Is There a Bulbar Pacemaker Responsible for the Ultradian Rhythm of Paradoxical Sleep?" *Archives Italiennes de Biologie* 134 (1995): 39–56.

[21]Depressed patients undergoing a combined treatment with an antidepressant drug and sleep deprivation showed a progressive decline of slow-wave activity, which due to the suppression of REM sleep was little interrupted (Van den Hoofdakker and Beersma, unpublished data). The data could be simulated by leaving the REM sleep process in its inactive state. See P. Achermann and A. A. Borbély, "Simulation of Human Sleep: Ultradian Dynamics of EEG Slow-Wave Activity," *Journal of Biological Rhythms* 5 (1990): 141–157.

[22]Brezinová, "Sleep Cycle Content and Sleep Cycle Duration"; Schulz and Tetzlaff, "Distribution of REM Latencies"; Fukuda, Miyashita, and Inugami, "Sleep Onset REM Periods Observed"; Miyashita, Fukuda, and Inugami, "Effects of Sleep Interruption on REM-NREM Cycle."

[23]H. W. Agnew, W. B. Webb, and R. L. Williams, "Comparison of Stage Four and 1-REM Sleep Deprivation," *Perceptual and Motor Skills* 24 (1967): 851–858; W. Dement, "The Effects of Dream Deprivation," *Science* 131 (1960): 1705–1707; A. Kales et al., "Dream Deprivation: An Experimental Reappraisal," *Nature* 204 (1964): 1337–1338; H. Sampson, "Deprivation of Dreaming Sleep by Two Methods. I. Compensatory REM Time," *Archives of General Psychiatry* 13 (1965): 79–86.

[24]T. Endo et al., "REM Sleep Regulation in Humans: Effects of Selective REM Sleep Deprivation," *Journal of Sleep Research* 5 (1996): 60; T. Endo et al., "Selective REM Sleep Deprivation in Humans: Effects on Sleep and Sleep EEG," *American Journal of Physiology* (forthcoming).

[25]Rats with a lesioned SCN still exhibit a rebound after sleep deprivation. See R. E. Mistlberger et al., "Recovery Sleep Following Sleep Deprivation in Intact and Suprachiasmatic Nuclei Lesioned Rats," *Sleep* 6 (1983): 217–233; I. Tobler, A. A. Borbély, and G. Groos, "The Effect of Sleep Deprivation on Sleep in Rats with Suprachiasmatic Lesions," *Neuroscience Letters* 42 (1983): 49–54; and L. Trachsel et al., "Sleep Homeostasis in Suprachiasmatic Nuclei-Lesioned Rats: Effects of Sleep Deprivation and Triazolam Administration," *Brain Research* 589 (1992): 253–261.

[26]D.-J. Dijk and C. A. Czeisler, "Contribution of the Circadian Pacemaker and the Sleep Homeostat to Sleep Propensity, Sleep Structure and Electroencephalographic Slow-Waves and Sleep Spindle Activity in Humans," *Journal of Sleep Research* 2 (1993): 121–129.

[27]Ray Meddis, A. J. D. Pearson, and G. Lanford, "An Extreme Case of Healthy Insomnia," *Electroencephalography and Clinical Neurophysiology* 35 (1973): 213–214.

[28]D. Aeschbach et al., "Homeostatic Sleep Regulation in Habitual Short Sleepers and Long Sleepers," *American Journal of Physiology* 39 (1996): R41–R53.

[29]S. Coren, *Sleep Thieves* (New York: The Free Press, 1996).

[30]I. Tobler, "Is Sleep Fundamentally Different between Mammalian Species?" *Behavioral Brain Research* 69 (1995): 35–41.

[31]I. Tobler, P. Franken, and K. Jaggi, "Vigilance States, EEG Spectra and Cortical Temperature in the Guinea Pig," *American Journal of Physiology* 264 (1993): R1125–R1132.

[32]S. Daan, B. M. Barnes, and A. M. Strijkstra, "Warming up for Sleep? Ground Squirrels Sleep During Arousals from Hibernation," *Neuroscience Letters* 128 (1991): 265–268; L. Trachsel, D. M. Edgar, and H. C. Heller, "Are Ground Squirrels Sleep Deprived during Hibernation?" *American Journal of Physiology* 260 (1991): R1123–R1129; and T. Deboer and I. Tober, "Natural Hypothermia and Sleep Deprivation: Common Effects on Recovery Sleep in the Djungarian Hamster," *American Journal of Physiology* 271 (1996): R1364–R1371.

[33]I. Tobler, "Phylogenèse du sommeil," in O. Benoit and M. Foret, eds., *Le sommeil humain: bases expérimentales, physiologiques et physiopathologiques* (Paris: Masson, 1995), 3–16.

[34]I. Tobler and M. Neuner-Jehle, "24-h Variation of Vigilance in the Cockroach *Blaberus giganteus*," *Journal of Sleep Research* 1 (1992): 231–239; I. Tobler and J. Stalder, "Rest in the Scorpion: A Sleep-like State?" *Journal of Comparative Physiology A* 163 (1988): 227–235.

[35]M. Hanagasioglu and A. A. Borbély, "Effect of Voluntary Motor Activity on Sleep in the Rat," *Behavioral Brain Research* 4 (1982): 359–368.

[36]D.-J. Dijk and C. A. Czeisler, "Body Temperature is Elevated during the Rebound of Slow-Wave Sleep following 40-h of Sleep Deprivation on a Constant Routine," *Journal of Sleep Research* 2 (1993): 117–120.

[37]J. Horne, *Why We Sleep: The Functions of Sleep in Humans and Other Mammals* (Oxford, England: Oxford University Press, 1988); H. S. Driver et al., "Prolonged Endurance Exercise and Sleep Disruption," *Medicine and Science in Sports and Exercise* 26 (1994): 903–907.

[38]Achermann et al., "A Model of Human Sleep Homeostasis Based on EEG Slow-Wave Activity."

[39]D.-J. Dijk et al., "Quantitative Analysis of the Effects of Slow Wave Sleep Deprivation during the First 3h of Sleep on Subsequent EEG Power Density," *European Archives of Psychiatry and Clinical Neuroscience* 236 (1987): 323–328; D.-J. Dijk and D. G. M. Beersma, "Effects of SWS Deprivation on Subsequent EEG Power Density and Spontaneous Sleep Duration," *Electroencephalography and Clinical Neurophysiology* 72 (1989): 312–320.

[40]L. M. Mukhametov, A. Y. Supin, and I. G. Polyakova, "Interhemispheric Asymmetry of the Electroencephalographic Sleep Patterns in Dolphins," *Brain Research* 134 (1977): 581–584.

[41]A. I. Oleksenko et al., "Unihemispheric Sleep Deprivation in Bottlenose Dolphins," *Journal of Sleep Research* 1 (1992): 40–44.

[42]I. N. Pigarev, H.-C. Nothdurft, and S. Kastner, "Evidence for Asynchronous Development of Sleep in Cortical Areas," *NeuroReport* 8 (1997): 2557–2560.

[43]H. Kattler, D.-J. Dijk, and A. A. Borbély, "Effect of Unilateral Somatosensory Stimulation Prior to Sleep on the Sleep EEG in Humans," *Journal of Sleep Research* 3 (1994): 159–164.

[44]J. M. Krueger and F. Obál, "A Neuronal Group Theory of Sleep Function," *Journal of Sleep Research* 2 (1993): 63–69.

[45]Y. Harrison and J. A. Horne, "Sleep Loss Affects Frontal Lobe Function as Shown in Complex 'Real World' Tasks," *Sleep Research* 25 (467) (1996); Y. Harrison and J. A. Horne, "Sleep Deprivation Affects Speech," *Sleep* 20 (1997): 871–877.

[46]J. A. Horne, "Human Sleep, Sleep Loss and Behavior: Implications for the Prefrontal Cortex and Psychiatric Disorder," *British Journal of Psychiatry* 162 (1993): 413–419.

[47]E. Werth, P. Achermann, and A. A. Borbély, "Brain Topography of the Human Sleep EEG: Antero-Posterior Shifts of Spectral Power," *NeuroReport* 8 (1996): 123–127; E. Werth, P. Achermann, and A. A. Borbély, "Fronto-Occipital EEG Power Gradients in Human Sleep," *Journal of Sleep Research* 6 (1997): 102–112.

[48]Recent studies of regional cerebral blood flow by PET showed that slow-wave sleep is associated with selective deactivation of heteromodal association areas in the frontal and parietal lobes. A. R. Braun et al., "Regional Cerebral Blood Flow throughout the Sleep-Wake Cycle, An $H_2^{15}O$ PET Study," *Brain* 120 (1997): 1173–1197.

[49]G. Moruzzi, "The Functional Significance of Sleep with Particular Regard to the Brain's Mechanisms Underlying Consciousness," in J. C. Eccles, ed., *Brain and Conscious Experience* (New York: Springer, 1966), 345–388.

[50]D. K. Welsh et al., "Individual Neurons Dissociated from Rat Suprachiasmatic Nucleus Express Independently Phased Circadian Firing Rhythms," *Neuron* 14 (1995): 697–706.

[51]D. H. Hubel, "Single Unit Activity in Striate Cortex of Unrestrained Cats," *Journal of Physiology* 147 (1959): 226–238; and E. V. Evarts, "Effects of Sleep and Waking on Spontaneous and Evoked Discharge of Single Units in Visual Cortex," *Federation Proceedings* 19 (1960): 828–837.

[52]M. Steriade and J. A. Hobson, "Neuronal Activity during the Sleep-Waking Cycle," *Progress in Neurobiology* 6 (1976): 155–376; M. Steriade, E. G. Jones, and R. R. Llinás, *Thalamic Oscillations and Signaling* (New York: Wiley, 1990); M. Steriade, D. A. McCormick, and T. J. Sejnowski, "Thalamocortical Oscillations in the Sleeping and Aroused Brain," *Science* 262 (1993): 679–685; M. Steriade, D. Contreras, and F. Amzica, "Synchronized Sleep Oscillations and Their Paroxysmal Developments," *Trends in Neurosciences* 17 (1994): 199–208; and D. A. McCormick and T. Bal, "Sleep and Arousal: Thalamocortical Mechanisms," *Annual Review of Neuroscience* 20 (1997): 185–215.

[53]M. Steriade, A. Nuñez, and F. Amzica, "A Novel Low (<1 Hz) Oscillation of Neurocortical Neurons in Vivo: Depolarizing and Hyperpolarizing Components," *Journal of Neuroscience* 13 (1993): 3252–3265; M. Steriade et al., "The Slow (<1 Hz) Oscillation in Reticular Thalamic and Thalamocortical

Neurons: Scenario of Sleep Rhythm Generation in Interacting Thalamic and Neocortical Networks," *Journal of Neuroscience* 13 (1993): 3284–3299; P. Achermann and A. A. Borbély, "Low-Frequency (<1 Hz) Oscillations in the Human Sleep EEG," *Neuroscience* 81 (1997): 213–222; and F. Amzica and M. Steriade, "The K-complex: Its Slow (<1 Hz) Rhythmicity and Relation with Delta Waves," *Neurology* 49 (1997): 952–959.

[54]M. Pompeiano, C. Cirelli, and G. Tononi, "Effects of Sleep Deprivation on Fos-like Immunoreactivity in the Rat Brain," *Archives Italiennes de Biologie* 130 (1992): 325–335; M. Pompeiano, C. Cirelli, and G. Tononi, "Immediate-Early Genes in Spontaneous Wakefulness and Sleep: Expression of c-Fos and NGFI-A mRNA and Protein," *Journal of Sleep Research* 3 (1994): 80–96; C. Cirelli, M. Pompeiano, and G. Tononi, "Sleep Deprivation and c-Fos Expression in the Rat Brain," *Journal of Sleep Research* 4 (1995): 92–106.

[55]C. Cirelli, M. Pompeiano, and G. Tononi, "Neuronal Gene Expression in the Waking State: A Role for the Locus Coeruleus," *Science* 274 (1996): 1211–1215; C. Cirelli and G. Tononi, "Changes in Anti-phosphoserine and Anti-phosphothreonine Antibody Binding during the Sleep-Waking Cycle and after Lesions of the Locus Coeruleus," *Sleep Research Online* 1 (1998): 11–18.

[56]R. D. Hawkins, E. R. Kandel, and S. A. Siegelbaum, "Learning to Modulate Transmitter Release: Themes and Variations in Synaptic Plasticity," *Annual Review of Neuroscience* 16 (1993): 625–665; R. Bourtchuladze et al., "Deficient Long-Term Memory in Mice with a Targeted Mutation of the Camp-Responsive Element Binding Protein," *Cell* 79 (1994): 59–68; J. C. Yin et al., "CREB as a Memory Modulator: Induced Expression of a dCREB2 Active Isoform Enhances Long-Term Memory in Drosophila," *Cell* 81 (1995): 107–115. In a few cases, the need for immediate early gene expression for the occurrence of long-term plastic changes has been directly demonstrated. Cf. C. M. Alberini et al., "C/EBP is an Immediate-Early Gene Required for the Consolidation of Long-Term Facilitation in Aplysia," *Cell* 76 (1994): 1099–1114.

[57]C. Cirelli, M. Pompeiano, and G. Tononi, "Neuronal Gene Expression in the Waking State: A Role for the Locus Coeruleus," *Science* 274 (1996): 1211–1215.

[58]T. Kasamatsu, "Adrenergic Regulation of Visuocortical Plasticity: A Role of the Locus Coeruleus System," *Progress in Brain Research* 88 (1991): 599–616; Q. Gu and W. Singer, "Involvement of Serotonin in Developmental Plasticity of Kitten Visual Cortex," *European Journal of Neuroscience* 7 (1995): 1146–1153; M. F. Bear and W. Singer, "Modulation of Visual Cortical Plasticity by Acetylcholine and Noradrenaline," *Nature* 320 (1986): 172–176.

[59]Direct confirmation that the firing of these systems controls the amount of neuromodulators in a local volume of brain tissue has recently been obtained. F. M. Florin-Lechner et al., "Electrical Stimulation of the Locus Coeruleus (LC): Effect of Frequency and Pattern of Discharge on Extracellular Norepineprine (NE) Release in the Prefrontal Cortex (PFC)," *Society for Neuroscience Abstracts* 22 (1996): 240.6; and C. W. Berridge et al., "Relationship between Locus Coeruleus Neuronal Activity and Norepinephrine Release in Neocortex," *Society for Neuroscience Abstracts* 22 (1996): 240.3.

[60]For example, S. L. Foote and J. H. Morrison, "Extrathalamic Modulation of Cortical Function," *Annual Review of Neuroscience* 10 (1987): 67–95; T.

Kasamatsu, "Adrenergic Regulation of Visuocortical Plasticity: A Role of the Locus Coeruleus System," *Progress in Brain Research* 88 (1991): 599–616.

[61]Moruzzi, "The Functional Significance of Sleep with Particular Regard to the Brain's Mechanisms Underlying Consciousness," in Eccles, ed., *Brain and Conscious Experience*, 345–388; G. Moruzzi, "The Sleep-Waking Cycle," *Ergebnisse der Physiologie* 64 (1972): 1–165.

[62]Ibid.

[63]T. A. Rhyner, A. A. Borbély, and J. Mallet, "Molecular Cloning of Forebrain mRNAs which are Modulated by Sleep Deprivation," *European Journal of Neuroscience* 2 (1990): 1063–1073; M. Pompeiano, C. Cirelli, and G. Tononi, "Reverse Transcription mRNA Differential Display: A Systematic Molecular Approach to Identify Changes in Gene Expression across the Sleep-Waking Cycle," in Ralph Lydic, ed., *Molecular Regulation of Arousal States* (Boca Raton, Fla.: CRC Press, 1998), 157–166; C. Cirelli and G. Tononi, "Differences in Gene Expression Between Sleep and Waking as Revealed by RNA Differential Display," *Molecular Brain Research* (forthcoming).

[64]J. H. Benington and H. C. Heller, "Restoration of Brain Energy Metabolism as the Function of Sleep," *Progress in Neurobiology* 45 (1995): 347–360.

[65]P. Maquet et al., "Cerebral Glucose Utilization During Sleep-Wake Cycle in Man Determined by Positron Emission Tomography and [18F]2-Fluoro-2-Deoxy-D-Glucose Method," *Brain Research* 513 (1990): 136–143; C. Kennedy et al., "Local Cerebral Glucose Utilization in Non–Rapid Eye Movement Sleep," *Nature* 297 (5864) (1982): 325–327; E. Ravussin et al., "Determinants of 24-Hour Energy Expenditure in Man: Methods and Results using a Respiratory Chamber," *Journal of Clinical Investigation* 78 (1986): 1568–1578.

[66]A recent microdialysis study has reported increased extracellular concentration of several amino acids in the cat thalamus during non-REM sleep. K. A. Kékesi et al., "Slow Wave Sleep is Accompanied by Release of Certain Amino Acids in the Thalamus of Cats," *NeuroReport* 8 (1997): 1183–1186. Another report has shown an increase in glutamate content in the rat cerebral cortex after REM sleep deprivation. L. Bettendorff et al., "Paradoxical Sleep Deprivation Increases the Content of Glutamate and Glutamine in Rat Cerebral Cortex," *Sleep* 19 (1996): 65–71. Clearly, further studies are needed to establish whether glutamate release in the cortex changes during the sleep-waking cycle and after sleep deprivation.

[67]D. Contreras, I. Timofeev, and M. Steriade, "Mechanisms of Long-Lasting Hyperpolarizations underlying Slow Sleep Oscillations in Cat Corticothalamic Networks," *Journal of Physiology* 494 (1996): 251–264; M. Steriade, "Basic Mechanisms of Sleep Generation," *Neurology* 42, suppl. 6 (1992): 9–17.

[68]G. Tononi and C. Cirelli, "Changes in the Expression of Mitochondrial Genes during the Sleep-Waking Cycle," *Society for Neuroscience Abstracts* (1997): 1846; Cirelli and Tononi, "Differences in Gene Expression between Sleep and Waking as Revealed by RNA Differential Display."

[69]H. Zepelin, "Mammalian Sleep," in M. H. Kryger, T. Roth, and W. C. Dement, eds., *Principles and Practice of Sleep Medicine* (Philadelphia, Pa.: Saunders Company, 1994), 69–80.

[70]F. Crick and G. Mitchison, "REM Sleep and Neural Nets," *Behavioral Brain Research* 69 (1995): 147–155; A. Giuditta et al., "The Sequential Hypothesis of the Function of Sleep," *Behavioral Brain Research* 69 (1995): 157–166; E. Hennevin et al., "Processing of Learned Information in Paradoxical Sleep: Relevance for Memory," *Behavioral Brain Research* 69 (1995): 125–135.

[71]G. A. Marks et al., "A Functional Role for REM Sleep in Brain Maturation," *Behavioral Brain Research* 69 (1995): 1–11.

[72]M. Jouvet, "Paradoxical Sleep and the Nature-Nurture Controversy," *Progress in Brain Research* 53 (1980): 331–346; M. Jouvet, "Le sommeil paradoxal: Est-il le gardien de l'individuation psychologique?" *Canadian Journal of Psychology* 45 (1991): 148–168.

[73]J. L. Kavanau, "Sleep and Dynamic Stabilization of Neural Circuitry: A Review and Synthesis," *Behavioral Brain Research* 63 (1994): 111–126; J. M. Krueger et al., "Brain Organization and Sleep Function," *Behavioral Brain Research* 69 (1995): 177–185.

[74]C. H. Bailey, D. Bartsch, and E. R. Kandel, "Toward a Molecular Definition of Long-Term Memory Storage," *Proceedings of the National Academy of Science* 93 (1996): 13445–13452.

[75]R. D. Fields, "Signaling from Neural Impulses to Genes," *The Neuroscientist* 2 (1996): 315–325.

[76]M. A. Wilson and B. L. McNaughton, "Reactivation of Hippocampal Ensemble Memories during Sleep," *Science* 265 (1994): 676–679.

George L. Gabor Miklos

The Evolution and Modification of Brains and Sensory Systems

T HE TRUE TEST OF THE ASSERTION that we understand how brains develop and change as a result of experience, that we know how evolutionary novelties of brains and sensory systems evolve, is whether we are able to *predict* the consequences of natural variation in populations at the molecular, cellular, neuronal, and organismal levels, and to *predict* the consequences of genomic perturbations that ultimately lead to complex brain disorders. If we cannot reach these goals, brain research will become an increasingly heterogeneous collection of historical "just-so" stories. If, however, robust predictive frameworks are developed about brain structure and function, then the evolution of brain complexity, and an understanding of brain disorders, will emerge as a major cornerstone of biology. To put this task into perspective, one needs only to pose the following questions: What are the limits to existing neuroanatomies, sensory systems, and behaviors? How far can we move beyond understanding what evolutionary processes *have* produced, and instead modify nervous and sensory systems to determine what *can be* produced? How useful are computer analogies to an understanding of brain functions? This essay examines these issues.

GENOMES, NEURONS, AND EVOLUTION

There are basically three experimental avenues that are undertaken to understand the development and workings of brains

George L. Gabor Miklos is Senior Fellow at The Neurosciences Institute.

and sensory systems. The first, and broadest, is the in-depth study of each and every organism in the phylogenetic spectrum, including an analysis of its developmental, morphological, neural, and behavioral complexities as it constructs its particular world. The second is the focused study of the human condition, in which "model" organisms are utilized to aid in understanding human disorders. The third, and newest avenue, is a fusion of synthetic modeling and the use of automata, an approach that attempts to determine the principles of nervous system structure and function as transitions are made from one level to another.[1]

In addressing the challenges associated with each of these endeavors, one can begin by asking how much progress has been made and how distant are the goals that still need to be attained. I begin here with the bottom-up genomic approach, and then examine the variation in neuroanatomies and sensory systems between individuals of the same species, as well as variation across the evolutionary spectrum.

Genomic Engine Rooms

The various worldwide genome projects and their ancillary technological counterparts, the transcriptome and proteome projects, are yielding unambiguous data on the molecular diversity that represents the evolutionary end products of nearly four billion years of variation, selection, and historical accidents.[2] These high-throughput "robogenomics" approaches are highly successful. The genomic sequences of nearly a dozen bacteria are finished, the complete sequence of a yeast is published, and that of a worm is nearly complete.[3] There is confidence that the human genome will be available within the next decade—with that of a fly and a plant (the wall cress) within the next few years—and that in the postsequence era a transition can be made to the broad new field of epigenomics.

The successes of medical diagnostics, molecular medicine, and mammalian cloning, and the manner in which most human phenotypes are shoehorned into the "one gene–one disease" category, have reinforced an essentially linear, mechanical, and reductionist view of biology at its first approximation. Most animals, but particularly the invertebrates, are considered to be machine-like, with brains that are little more than circuit boards and molecular switches, cogs, and ratchets providing the under-pinnings of pathways involved in cellular differentiation. Building a brain appears to be only a slight extension of understanding the processes of cellular differentiation. In essence, variation, epigenesis, thresholds, nonlinearities, emergent properties, and the "levels" problem are generally considered to be irritants of an almost semantic nature, none of which are thought to be of importance for the mainstream task of simply gathering more data. In fact, some believe that the notion that so-called emergent properties are a prerequisite for understanding living organisms is "a bunch of yak, all talk, and nothing more."[4] A common assumption is that the reductionist approaches will inexorably allow us to understand not only how brain development occurs but, ultimately, how organisms behave. However, excellent data sets already exist showing that the difficult tasks actually lie in utilizing the evolutionary, molecular, and synthetic modeling approaches to understand how brain networks arise and function in topobiological contexts.[5]

"Information" and Computer Analogies

In the neurosciences, the dominant parallelism to this machine-like view of development is the computer-like view of nervous systems. For invertebrate animals in particular, nervous systems are still generally thought of as a combination of circuit boards, hard-wired during embryogenesis, with stereotyped networks that analyze, store, and process "information"; the organism responds to the "environment" largely as an inflexible robot. However, it is becoming clear that even apparently simple insect brains undergo extensive neuroanatomical changes as a result of experience. Furthermore, it is prudent to remember that "information" only emerges from an interaction between an organism and its environment; it does not exist in an immutable form in a biological world. It is also clear that what is simulated as computer learning and memory is *improving a search in a predefined body of knowledge*. In real organisms, learning and memory deal primarily with *expanding a knowledge base* under conditions of novelty.

To address the intertwined issues involving computer-like and organismal views of brains, we examine the variation and nonlinearities at the different levels of genes, neuroanatomies, and sensory systems in different organisms. A knowledge of the available biological building blocks, their properties, and their interactions during embryogenesis leads naturally to the issues of the evolution of brain complexity, the modification of brains via genetic engineering, and the desirability of the synthetic modeling of brain structure and function.

More Is Not Always Better

Since it is a widely held belief that more genes, more cell types, more neurons, more synapses, more axonal and dendritic arborizations, and more glial cells give rise to greater morphological and behavioral complexity, I will now examine data that bear on this popular assumption (table 1). I will demonstrate that, except in the extreme comparisons, there is little relationship between gene number, neural components, and the morphological or behavioral characteristics of organisms.

Table 1. *Approximate* Gene and Neuron Numbers in the Brains[b] or Entire Nervous Systems[c] of Organisms in Different Evolutionary Lineages (n = not available).

Organism	Genes	Neurons
worm	16,000	302[c]
fly	12,000	250,000[b]
miniature wasp	n	5,000[b]
honeybee	n	850,000[b]
marine snail	n	20,000[c]
octopus	n	520 million[c]
puffer fish	70,000	n
miniature salamander	n	300,000[b]
laboratory mouse	70,000	40 million[b]
human being	70,000	85,000 million[b]
whale and elephant	70,000	200,000 million[b]
tobacco plant	43,000	0

Source: Miklos, "Molecules and Cognition"; George L. G. Miklos and Gerald M. Rubin, "The Role of the Genome Project in Determining Gene Function: Insights from Model Organisms," *Cell* 86 (1996): 521–529; Sydney Brenner et al., "Characterization of the Puffer Fish (*Fugu*) Genome as a Compact Model Vertebrate Genome," *Nature* 366 (1993): 265–268; Gerhard Roth et al., "Cell Size Predicts Morphological Complexity in the Brains of Frogs and Salamanders," *Proceedings of the National Academy of Sciences* 91 (1994): 4796–4800; and Martin W. Simmen et al., "Gene Number in an Invertebrate Chordate," *Proceedings of the National Academy of Sciences* 95 (1998): 4437–4440.

The comparison between the mouse and the octopus is particularly appropriate as an exemplar of the problem of the development and evolution of biological complexity, and of the coevolution of brains and sensory systems. An average adult octopus has a multilobed, richly interconnected brain of approximately 150 million neurons, with an additional 370 million neurons organized in ganglia at the bases of its arms.[6] These arms can lengthen, shorten, twist, or bend at any point and in any direction, and the animal can carry out tasks such as building a shelter and manipulating objects. These arms have virtually unlimited degrees of freedom and pose special problems, not only in terms of the coevolution of neural controls of sensory structures in the absence of joints as reference points, but also in terms of the evolution of novel behaviors in novel environments.

The mouse also has a richly interconnected brain, with approximately 40 million neurons (less than the number found at the base of one arm of the octopus); it has evolved a totally different neuroanatomy and has developed different behavioral attributes. These two organisms are poles apart at most biological levels. However, it is presently not possible to objectively decide which of them is more complex at the genomic, cellular, developmental, neuroanatomical, morphological, or behavioral levels. The reason is simple: we have as yet no predictive theory of biological complexity that transcends different biological levels and incorporates changes in complexity through both developmental and evolutionary time.

A second example again involves a comparison between an invertebrate and a vertebrate. The approximately 300,000-neuron brain of a miniature salamander has less neurons than the 850,000 neurons in the brain of the worker honeybee, and their neuroanatomies and behavioral characteristics are very different. In addition, different species within this salamander family can have different neuron numbers and similar behavioral characteristics or, alternatively, similar neuron numbers and different behavioral characteristics. For example, those salamanders with the lowest anatomical brain complexity (as measured by the number of distinct nuclei and multiply laminated structures) are the arboreal species, and these in fact are the most acrobatic. However, their apparent neuroanatomical simplicity does not correlate with their apparent behavioral complexity. This is in line with previous studies on relative brain size and behavioral complexity, especially within mammals and within social insects. So far, none of these studies has provided a foundation for a general measure of behavioral complexity, although the concept of neural complexity now has a significant foundation.[7]

However, one does not need to make such distant evolutionary comparisons. Among insects, the minute brains of some parasitic wasps probably contain less than 5,000 neurons, since the head size of some of these organisms is on the order of 30 microns, about the diameter of some single human cells. These wasps have a life cycle that appears no less complex than that of some of their larger relatives. Furthermore, among mollusks, the nervous systems of the miniature ones that live *between* sand grains and have body lengths of 500 microns or so are unlikely to even approach the 20,000-neuron figure of a commonly used marine snail. Except for the extreme case of the predatory squid and octopus, one is hard-pressed to find a significant correlation between neuronal number and behavior in mollusks.

The numerology at the genomic level in different phyla is similarly not correlated with apparent morphological complexity. The number of genes in a small worm and a fly are approximately the same, about 12,000–16,000, a figure that is little different from that of a unicellular protozoan and well below the estimated numbers in some plant species.[8] In vertebrates, gene number from bony fish to humans is roughly around 70,000, yet the morphologies and behavioral characteristics of the extant members are exceedingly diverse (electroreceptive fish, snakes with infrared vision, echolocating bats and whales, and humans with mathematical skills). Clearly, the variation in morphogenesis and nervous system evolution from the time of bony fish to humans has occurred in the presence of near-constant gene number. This is a most important nonlinearity, anchored at one end largely with the molecular data from the puffer fish and at the other with extensive database information from mice and humans.

Once again, one does not have to resort to broad evolutionary comparisons; they can be restricted to mammals, which themselves show extensive morphological and behavioral variation, but where gross neuroanatomies are comparable. For example, not only are bat and rat neuroanatomies similar, but over 95 percent of the nuclei in the rat brain are found in the human brain. Rodent and human brains are also alike in terms of the distribution and abundance of neurotransmitters. Furthermore, neuron numbers in miniature primates such as pygmy marmosets (which are less than 4 inches in size, weigh only about 6 ounces, and have a brain volume of only about 1.5 cubic centimeters) must be at least two orders of magnitude less than that of chimpanzees, but the behavioral characteristics of these primates are broadly comparable.

The ultimate primate comparison in nonlinearities is found between humans and chimpanzees. The similarities in gene structure, embryology, and neuroanatomy are sometimes almost indistinguishable, yet the evolution of language has had such large effects on human evolution that its consequences far exceed those of any morphological or behavioral changes in the entire vertebrate lineage. Owing to the huge genomic, embryological, and neuroanatomical overlap between chimpanzees and human beings, the evolution of these nonlinearities must have involved few (if any) new genes.

The foregoing data unambiguously indicate that no simple relationships exist between gene and neuron number and apparent morphological and behavioral complexity, and the numbers themselves cannot be used in a predictive sense to come to terms with the behavioral complexities of organisms in different evolutionary lineages. What then can be learned from other available data on the *rates* at which complex organisms can evolve?

Evolutionary Time Windows

A major complexity transition from unicellular to multicellular forms is seen in the fossil record at about 530 million years ago, and its arrival is clearly marked by the remains of a heterogeneous group of organisms known as the small shelly fauna, before which there is no fossil evidence for organisms with complex nervous systems. Less than 5 million years later, organisms such as trilobites and various other predatory animals appear abruptly, already possessing compound eyes and various chemo-, mechano-, and proprio-sensory systems. It is clear that sophisticated neuromuscular and sensory systems were assembled and "matched" in a remarkably short period of evolutionary time.[9]

In the case of vertebrates, the fossil record is equally informative. It reveals that the evolution of brains and sensory systems was rapid, since eyes, otic, and lateral line systems were fully operational by 480 million years ago. The gross neuroanatomy of these early vertebrates, as exemplified by that of the extinct jawless fish, is very similar to that of modern lampreys and structurally similar to that of craniate vertebrates in general.[10] The brain cavity of the earliest jawless fish already had five major divisions: telencephalon, diencephalon, mesencephalon, metencephalon, and myelencephalon. These fish also already possessed the exit foramina for the ten cranial nerves, the olfactory capsules, the orbits, and the labyrinth cavities. Furthermore, the first jawed fish (the Osteolepiforms) had brain cavities that were essentially of the same type as the majority of the transitional amphibians (the Stegocephalians), frogs, and reptiles. In summary, fish brains evolved rapidly somewhere between 515 and 480 million years ago and were utilized largely unaltered by amphibians and reptiles; their fundamental gross neuronal architecture has been retained over a 500-million-year period.

These evolutionary data are important. They not only illustrate the rapidity with which biological systems *actually* underwent real evolutionary transitions, but they illustrate a general principle that occurs at many levels: the faster a transition occurs, the lower the probability that the potential universe of neuronal or morphogenetic space is able to be explored before it is locked into place in an embryogenic context. In short, it is probable that not all brains and sensory systems that *could* have evolved actually did so.

Given this overview of gene and neuron number in different evolutionary lineages, and the rates at which complex brains and sensory systems evolved at their very origins, I now turn to the fundamental issue of variation in biological systems. These data sets on variation impinge upon two diametrically opposed views of brain development, function, and evolution. One involves the organismal panorama of *no instruction—variation—selection*, leading to the adaptation of an organism to an initially unla-

beled world. Another, the *instructionist computer-like* view, requires extensive prespecification and does not as yet operate
under the stringent constraints of ongoing biological variability
in both somatic and evolutionary time.

Variation and Genetic Backgrounds

For the purposes of comparisons between distantly related organisms, I dealt with neuronal number in an organism as if it
were a constant. In fact, variation between individuals, and
variation within an individual in somatic time, can be great.

I begin by first considering the small invertebrates, where the
conventional wisdom has long been that their nervous systems
are hard-wired, their behaviors are highly stereotyped, and the
organisms themselves are akin to mini-robots. The data on these
issues (as opposed to the folklore) point to a quite different state
of affairs. Careful neuroanatomical and genetical analyses have
uncovered remarkable structural plasticity in the fly, revealing
that the brain is not only highly variable in size but that most
regions, including the optic lobes, the corpora pedunculata, the
central complex, and the central brain, are *continuously reorganized during somatic time depending on the specific living conditions with which the fly is faced.*[11] When flies live in large-
population cages with various odors and plants, they have approximately 15 percent more Kenyon cells in their major integrative brain center (the corpora pedunculata) than do flies that
are kept as isolated individuals. When kept in pairs, the sizes of
different parts of the brain are differentially affected depending
on the sex of the partner. Larvae grown under crowded conditions give rise to adults that have larger corpora pedunculata
than those grown in uncrowded conditions, and the female is
more affected than the male. Furthermore, there is experience-
dependent developmental plasticity in the optic lobes, with visual stimulation modifying the size of this structure. It seems
that all, or most, parts of the fly brain are variable and that
different experiences cause differential changes in different parts
of the brain. The situation is no different in natural populations
of the worker honeybee. It is clear that experience-dependent
plasticity is not an exclusive characteristic of vertebrates.

An additional important finding comes from genetics: *brain structure is different in different genetical backgrounds*. In the particular case of a mutant known as the *mushroom body miniature* gene in the fly, the anatomical phenotype of the brain can vary from near-normalcy to being totally mutant, depending on the genetical background in which the same mutation is placed.[12] While the generality of this *genetical background-phenotype* influence is not new for either flies or mice, it is beginning to attract more attention since the implications for between individual variation are so important. In mice, for example, an analysis of the "knockout" of a particular gene, the epidermal growth factor receptor, reveals that this perturbation can either lead to death during embryogenesis in some strains of mice or, alternatively, the mice can live for many weeks when this *same* perturbation is placed in other strains.[13] Many other mutations in both mice and humans give rise to variable phenotypes within the same family or the same litter, even though different individuals are carrying the same genomic perturbation. For example, mutations in the breast cancer genes, known as BRCA1 and BRCA2, and in a colon cancer gene, known as APC, can exist in different individuals of the same family, yet some members develop cancers whereas others do not. It is clear that other than in exceptional situations such as clones, no two organisms (or no two children within the same family) will possess either identical genetical backgrounds or identical neuroanatomies.

Furthermore, in different strains of mice, the number of neurons varies greatly in such different regions of the brain as the hippocampus, olfactory bulb, cerebellum, locus coeruleus, substantia nigra, and neocortex, with the differences in any one region varying from approximately 25 to 100 percent. A detailed examination of thirty-one different inbred and outbred strains of mice as well as wild species, subspecies, and their various hybrids has revealed, for example, that the number of retinal ganglion cells varies from approximately 32,000 to 87,000.[14] In macaque monkeys and humans, the size of the cortex itself and the size of individual cytoarchitectonic areas are highly variable, with a range in neuronal number exceeding plus or minus 50 percent between individuals.[15] Neuronal number in the lateral geniculate nucleus of macaque monkeys varies from approximately 1 to 2 million between individuals; in addition, the densities of synaptic junctions also reveal considerable variation within and between individuals. The amount and distribution of neurotransmitters and receptors is also highly variable between individuals.

The within and between individual variation in the human cortex is also extensive. The coefficient of variation for human-brain volume is on the order of 10 percent, while variation for single architectonic areas in the striate and extra-striate cortex is on the order of 20 to 40 percent. In addition, the surface areas of the dorsolateral frontal cortex differ considerably in the brains of human monozygotic twins.[16] Finally, cortical surface areas that have been assigned to different functional modalities are not immutable in the somatic lifetime of an individual.

These results from flies, honeybees, mice, monkeys, and humans have significant implications for the extent to which information processing (connectionist) views of brain function are helpful to understanding organismal brain function.[17]

UNFORESEEN NOVELTIES AND BIOLOGICAL VARIATION

All the organismal data highlighted so far, but particularly that pertaining to variation between and within individuals, bears directly upon the computer-like view of nervous systems. In the computer-like view, a brain does not construct its world by *interacting* with it; rather, the environment stamps itself on the brain. In effect, "information" already exists as such in the environment and is processed by an organism that essentially acts as a biological filter. In order for the computer analogy to be useful in a predictive sense, it needs to be implemented in the equivalent of systems that are *variable in developmental and somatic time,* and in which connections are *continuously reorganized in somatic time* (just as occurs in real organisms). This is a challenging undertaking at best, since if one considers the developing visual cortex of the macaque monkey as a platform for computer-based analogies, it is estimated that 30,000 synapses are lost per second in the cortex during the period of sexual maturation.[18]

In addition, unlike the biological cases in which organisms hatch or are born into a world that is unlabeled—a world that is itself changing and cannot be prefigured—a computer system needs a completely prespecified data set of information about its world; without it, it is unable to function normally in the face of unforeseen novelties.[19] Such is not the case with organisms. Navigating honeybees foraging in novel environments, for example, use the pattern of polarized light as a compass; to our knowledge, they do not carry astronomical almanacs in order to look up all possible e-vector patterns.[20] Computers, on the other hand, cannot as yet function without instruction in a changing world in which unforeseen novelties arise. In a biological world, these novelties are not only internal ones of changing microanatomy; they can be other organisms whose interactions alter the evolutionary landscape.

The computer metaphor is also constrained by existing knowledge based on evolutionary and developmental data. The first data set in this arena is that which bears on the changes in internal neuroanatomy in a species over evolutionary time. The second data set is that derived from transgenic organisms (those that have had genetical material added to or deleted from their genomes). In the transgenic cases, the novel neuroanatomies and sensory systems that are genetically engineered into an organism are generated in a single embryogenesis in somatic time. The nearest formal equivalent to making these changes to a computer system would be to engineer massive changes in hardware but withhold programming changes. In most cases of transgenic modifications to brains and sensory systems of *both* invertebrates and vertebrates, the organism is viable and robust; at worst, it is able to limp along under laboratory conditions, owing to the degeneracy that is the mainstay of biological systems. However, the outcome of equivalent circuit-breaking alterations to computer systems is far more drastic, namely, a total shutdown.

Evolutionary Innovations

The natural evolutionary novelties that are well documented in the brains and sensory systems of different organisms are also relevant to the computer metaphor, particularly when one considers the consequences of innovations such as the evolution of echolocation in bats and whales, the evolution of the corpus callosum, the evolution of the jamming of bat radars by certain species of crickets, the evolution of infrared vision, and the evolution of electroreception in fish. In the coevolving brain and sensory systems of certain electric fish (the mormyrids, for example), part of the cerebellum has been enormously expanded; while these fish utilize two different brain pathways for electrolocation and communication, each with different receptor types, other electric fish, such as the gymnotids, use a single anatomically distinct receptor for both purposes. In these organisms, therefore, a complex phenotype involves not only the development of novel peripheral receptors (known as ampullary and tuberous forms), which are themselves morphologically and physiologically diverse, but also the modification of different brain nuclei. There is thus coevolution and "matching" of components both within and between organisms, which has allowed novel situations to be exploited without prespecification of these novel environments.

A second example of adaptation to an environment that is not prespecified is given by the evolution of infrared vision. In a classic example of degeneracy, infrared detection has evolved independently in both insects and reptiles, involving changes in both the sensory and nervous systems. Snakes such as vipers have facial pits, the membranes of which are warmed by infrared radiation. These pits effectively function as infrared eyes, and the signals are transferred via the nerve fibers of the trigeminal nerve to the modified optic tectum. In certain beetles of the family Buprestidae, evolution has produced paired pits adjacent to the bases of the middle legs; these pits are, in effect, infrared eyes, used in long-distance orientation towards forest fires.

Given these examples of morphological and neuroanatomical variation as it relates to the exploitation of novel environments, how do we address one of the questions that was posed at the beginning of this essay, namely, how do we move beyond understanding what evolutionary processes have *actually* produced and construct novel morphogenetic networks that will allow us to understand what *could*, and *can still*, evolve? To approach this question, we will briefly examine the impact of some transgenic technologies on organismal modifications, and the limits to which they are likely to be pushed.

TRANSGENIC MODIFICATIONS OF ORGANISMS

One of the most powerful methodologies for examining what has been conserved over different evolutionary lineages and what has altered is to move genes from one organism to another by transgenic means and examine the extent to which the components are able to maintain normal functions. The results are sometimes spectacular and other times quite mundane. For example:

- a particular protein from a fly can induce the formation of cartilage, bone, and bone marrow in a rat, even though all invertebrates are devoid of any such material;[21]

- a particular protein from a mouse, or similar proteins from worms and squid, can induce the formation of morphologically normal insect eyes; eyes can be made to appear not only in their usual position on the insect head but on many different parts of a fly, including its antennae and legs;[22]

- transgenic mice that lack synapsin, a major phosphoprotein thought to be indispensable to neurotransmitter release, show no apparent changes in well being or gross nervous system structure or function;[23]

- transgenic mice that lack complex gangliosides (sialic acid-containing glycosphingolipids), which are thought to be critical to neuritogenesis and synaptogenesis, show no major defects in gross behavior or nervous system structure and function.[24]

These results, a small sample from literally hundreds of successful transgenic modifications of organisms, indicate that genetic transfers (even between very different organisms) are a powerful means of modifying development in order to test hypotheses about brain function at many different levels. The transgenic technologies in both mice and flies have now moved beyond the levels at which single genes are replaced with those from a different organism to the level where whole developmental and neural architectures can be modified by a combination of changes in both gene and control elements.[25] Thus sophisticated systems for cutting and moving DNA in a genome are engineered into an organism, so that the organism itself does the cutting of its own genetical material upon a stimulus from the experimenter. In the fly, for example, these systems allow for the production of mutant or modified cell lines within the developing brain, as well as subsequent cell lineage and cell population interactions. In addition, without interfering with the developmental processes leading to adult brain structure, gene products from any organism or synthetic source can be expressed in specific parts of the fly nervous system and targeted to different subcellular locations such as synapses.

Interestingly, although the ability to modify organisms in exquisitely precise ways is becoming easier and easier to do, understanding how the transgenic changes are instantiated in a multi-layered system is becoming more and more difficult. The final measurement is ultimately at the level of behavior, with the intermediate morphogenetical steps being largely unknown. The major obstacle to understanding is that most organisms are "buffered" to differing extents against genomic alterations.[26] For example, in approximately 50 percent of the cases where a perturbation is made to any genome—be it that of a bacterium, yeast, worm, fly, fish, mouse, or human—it is difficult to determine if anything of significance has in fact happened to the organism; it appears normal, or near normal, under many laboratory conditions.

An additional imposing restriction to organismal manipulation is the size and combinatorial complexity of any system at a particular level; this becomes evident by highlighting what is required experimentally if one decides to investigate mammalian olfaction, for example, by manipulating its genomic components, which consist primarily of a huge multigene family. It is estimated that there are approximately one thousand different G protein–coupled olfactory receptors expressed in the olfactory epithelium in humans.[27] If the mammalian olfactory system is looked at holistically (from genes through neuroanatomy to behavior), we are presently unable to precisely modify the activities of hundreds of these proteins *simultaneously* by transgenic methodologies. Yet this is a minimum requirement if one is to begin to dissect this system at its most basic levels, namely, at the levels of the genome, the transcriptome, and the proteome. In addition to genetic engineering, one needs to measure small changes in phenotype that have effects on fitness, and this is a formidable task. In a biological sense, a pragmatic experimental plateau is reached when we move beyond a certain level and attempt to modify transgenically even a single cell.

This situation of dealing with populations of interacting molecules is really very similar to the current limitations seen in neurophysiology and in brain imaging. Just as there is an upper limit for the number of simultaneous transgenic manipulations that one can make even in a bacterium, let alone in a fly or a mouse, so also is there an upper limit to assaying populations of neurons with present technologies. For example, the number of neurons from which one can record intracellularly and simultaneously, compared to the potential number from which it is theoretically possible to record, is low. While intracellular and extracellular recordings allow small populations of cells to be analyzed, the trade-off is that the populational activity cannot be ascribed to individual synapses or neurons.

It is clear that there is a pressing need to exploit quite different approaches, ones in which the activities of populations of interacting components can be monitored and the principles of networks analyzed. This approach has been in existence for a decade and has been specifically applied to understanding nervous systems; it is termed synthetic neural modeling.[28] Importantly, in the automata and modeling used in these analyses, the environment, the phenotype, and the nervous system are all integrated, with selection acting on previously existing diversity in order to generate adaptive behavior in response to an initially unlabeled world.

PERSPECTIVES

It appears likely that the next century will contain a period of expanded transgenic biology, synthetic neural and embryogenic modeling, and the expanded use of biologically-based automata as a platform for understanding brain structure, function, and evolution. In the transgenic arena, there will be multiple modifications within metazoan genomes, including an increasing number of regulatory, rather than coding, sequences shuttled between different organisms. In addition, natural variation within and between species is likely to be much more extensively used to understand biological networks. The critical data that will be required in all these approaches center around nonlinear responses, thresholds, degeneracy, and, most importantly, fitness and phenotypic variation, all of which underpin embryogenesis and brain function, and each of which can only be analyzed partly by *in vivo* experimentation and partly by modeling. Finally, unraveling the underlying processes that contribute to human phenotypes—particularly neuropsychiatric ones, whose sheer complexity literally takes away one's breath[29]—will be one of the greatest challenges with which the human brain is confronted.

ACKNOWLEDGMENTS

I thank James F. Leckman for introducing me to the minutiae and overwhelming complexity of some human neuropsychiatric disorders, and David Edelman for his discussions on primate brain evolution. The author's research is supported by the Neurosciences Research Foundation.

ENDNOTES

[1]Gerald M. Edelman et al., "Synthetic Neural Modeling Applied to a Real-World Artifact," *Proceedings of the National Academy of Sciences* 89 (1992): 7267–7271; Gerald M. Edelman, *Neural Darwinism* (New York: Basic Books, 1987); and Gerald M. Edelman, "Neural Darwinism: Selection and Reentrant Signaling in Higher Brain Function," *Neuron* 10 (1993): 115–125.

[2]Victor E. Velculescu et al., "Characterization of the Yeast Transcriptome," *Cell* 88 (1997): 243–251; Marc R. Wilkins et al., "Current Challenges and Future Applications for Protein Maps and Post-Translational Vector Maps in Proteome Projects," *Electrophoresis* 17 (1995): 830–838; Richard Maleszka et al., "Data Transferability from Model Organisms to Human Beings: Insights from the Functional Genomics of the Flightless Region of *Drosophila*," *Proceedings of the National Academy of Sciences* 95 (1998): 3731–3736.

[3]Elizabeth Pennisi, "Microbial Genomes Come Tumbling In," *Science* 277 (5331) (1997): 1432. The common baker's yeast is *Saccharomyces cerevisiae*,

214   *George L. Gabor Miklos*

the nematode worm is *Caenorhabditis elegans*, the vinegar fly is *Drosophila melanogaster*, and the model laboratory plant is *Arabidopsis thaliana*.

[4]Sandra Blakeslee, "Some Biologists Ask 'Are Genes Everything?'" *New York Times*, 2 September 1997, science section, B7, B13.

[5]Gerald M. Edelman, *Topobiology* (New York: Basic Books, 1988); Gerald M. Edelman, *The Remembered Present: A Biological Theory of Consciousness* (New York: Basic Books, 1989); Gerald M. Edelman, *Bright Air, Brilliant Fire: On the Matter of the Mind* (New York: Basic Books, 1992); and George L. G. Miklos, "Molecules and Cognition: The Latterday Lessons of Levels, Language, and Lac," *Journal of Neurobiology* 24 (1993): 842–890.

[6]Miklos, "Molecules and Cognition."

[7]Giulio Tononi et al., "A Measure for Brain Complexity: Relating Functional Segregation and Integration in the Nervous System," *Proceedings of the National Academy of Science* 91 (1994): 5033–5037; and Giulio Tononi et al., "A Complexity Measure for the Selective Matching of Signals by the Brain," *Proceedings of the National Academy of Science* 93 (1996): 3422–3427.

[8]George L. G. Miklos and Gerald M. Rubin, "The Role of the Genome Project in Determining Gene Function: Insights from Model Organisms," *Cell* 86 (1996): 521–529.

[9]Miklos, "Molecules and Cognition"; George L. G. Miklos and Ken S. W. Campbell, "From Protein Domains to Extinct Phyla: Reverse-Engineering Approaches to the Evolution of Biological Complexities," in Stefan Bengtson, ed., *Early Life on Earth* (Nobel Symposium 84) (New York: Columbia University Press, 1994), 501–516; George L. G. Miklos, "Emergence of Organizational Complexities during Metazoan Evolution: Perspectives from Molecular Biology, Palaeonology and Neo-Darwinism," *Memoirs of the Australasian Association of Palaeontologists* 15 (1993): 7–41; and George L. G. Miklos et al., "The Rapid Emergence of Bio-electronic Novelty, Neuronal Architectures, and Organismal Performance," in Ralph J. Greenspan and Charalambos P. Kyriacou, eds., *Flexibility and Constraints in Behavioral Systems* (New York: John Wiley and Sons, Ltd., 1994), 269–293.

[10]Miklos et al., "The Rapid Emergence of Bio-electronic Novelty."

[11]Martin Heisenberg et al., "Structural Plasticity in the *Drosophila* Brain," *Journal of Neuroscience* 15 (3) (1995): 1951–1960; Martin Barth et al., "Experience-Dependent Developmental Plasticity in the Optic Lobe of *Drosophila melanogaster*," *Journal of Neuroscience* 17 (4) (1997): 1493–1504.

[12]J. Steven de Belle and Martin Heisenberg, "Expression of *Drosophila* Mushroom Body Mutations in Alternative Genetic Backgrounds: A Case Study of the Mushroom Body Miniature Gene (*mbm*)," *Proceedings of the National Academy of Science* 93 (1996): 9875–9880.

[13]David W. Threadgill et al., "Targeted Disruption of Mouse EGF Receptor: Effect of Genetic Background on Mutant Phenotype," *Science* 269 (1995): 230–238.

[14]Robert W. Williams et al., "Genetic and Environmental Control of Variation in Retinal Ganglion Cell Number in Mice," *Journal of Neuroscience* 16 (22) (1996): 7193–7205.

[15]Edelman, *Neural Darwinism*; Miklos, "Emergence of Organizational Complexities"; Williams et al., "Genetic and Environmental Control of Variation"; Robert W. Williams and Pasko Rakic, "Elimination of Neurons from the Rhesus Monkey's Lateral Geniculate Nucleus during Development," *Journal of Comparative Neurology* 272 (1988): 424–436; Pasko Rakic et al., "Synaptic Development of the Cerebral Cortex: Implications for Learning, Memory, and Mental Illness," *Progress in Brain Research* 102 (1994): 227–243; and Olaf Sporns, "Biological Variability and Brain Function," in John Cornwell, ed., *Binding the Mind: Consciousness and Human Identity* (New York: Oxford University Press, 1998).

[16]Grazyna Rajkowska and Patricia S. Goldman-Rakic, "Cytoarchitectonic Definition of Prefrontal Areas in the Normal Human Cortex: II. Variability in Locations of Areas 9 and 46 and Relationship to the Talairach Coordinate System," *Cerebral Cortex* 5 (1995): 323–337.

[17]Olaf Sporns, "Deconstructing Neural Constructivism," *Brain Behavioral Science* (forthcoming).

[18]Rakic et al., "Synaptic Development of the Cerebral Cortex."

[19]Claudia Carello et al., "Inadequacies of the Computer Metaphor," in Michael Gazzaniga, ed., *Handbook of Cognitive Neuroscience* (New York: Plenum Press, 1984), chap. 12, 229–247.

[20]Miklos et al., "The Rapid Emergence of Bio-electronic Novelty, Neuronal Architectures, and Organismal Performance."

[21]T. K. Sampath et al., "*Drosophila* Transforming Growth Factor B Superfamily Proteins Induce Endochronal Bone Formation in Mammals," *Proceedings of the National Academy of Science* 90 (1993): 6004–6008.

[22]William A. Harris, "Pax-6: Where to be Conserved is not Conservative," *Proceedings of the National Academy of Science* 94 (1997): 2098–2100.

[23]Thomas W. Rosahl et al., "Short Term Synaptic Plasticity is Altered in Mice Lacking Synapsin 1," *Cell* 75 (1993): 661–670.

[24]Kogo Takamiya et al., "Mice with Disrupted GM2/GD2 Synthase Gene Lack Complex Gangliosides but Exhibit Only Subtle Defects in their Nervous System," *Proceedings of the National Academy of Science* 93 (1996): 10662–10667.

[25]Andrea H. Brand and Norbert Perrimon, "Targeted Gene Expression as a Means of Altering Cell Fates and Generating Dominant Phenotypes," *Development* 118 (1993): 401–415; Jean-François Ferveur et al., "Genetic Feminization of Brain Structures and Changed Sexual Orientation in Male *Drosophila*," *Science* 267 (1995): 902–905; Ralph J. Greenspan, "Flies, Genes, Learning and Memory," *Neuron* 15 (1995): 747–750; Douglas A. Harrison and Norbert Perrimon, "Simple and Efficient Generation of Marked Clones in *Drosophila*," *Current Biology* 3 (1993): 424–433; Bruce A. Hay et al., "P Element Insertion-Dependent Gene Activation in the *Drosophila* Eye," *Proceedings of the National Academy of Science* 94 (1997): 5195–5200; Tian Xu and Gerald M. Rubin, "Analysis of Genetic Mosaics in Developing and Adult *Drosophila* Tissues," *Development* 117 (1993): 1223–1237.

[26]Edelman, *Neural Darwinism*; Miklos, "Molecules and Cognition"; Kathryn L. Crossin, "Functional Role of Cytotactin/Tenascin in Morphogenesis: A Modest Proposal," *Perspectives on Developmental Neurobiology* 2 (1994): 21–32; and Harold P. Erickson, "Gene Knockouts of *c-src*, Transforming Growth Factor b1, and Tenascin Suggest Superfluous, Nonfunctional Expression of Proteins," *The Journal of Cell Biology* 120 (1993): 1079–1081.

[27]Linda Buck, "Information Coding in the Vertebrate Olfactory System," *Annual Review of Neuroscience* 19 (1996): 517–544.

[28]Edelman et al., "Synthetic Neural Modeling Applied to a Real-World Artifact"; and Edelman, *Neural Darwinism*.

[29]James L. Leckman, "Pathogenesis of Tourette's Syndrome," *Journal of Child Psychology and Psychiatry* 38 (1997): 119–142.

Emilio Bizzi and Ferdinando A. Mussa-Ivaldi

The Acquisition of Motor Behavior

INTRODUCTION

RECENTLY, GREAT STRIDES HAVE BEEN MADE in understanding the neural foundations of motor behavior. Through the combined efforts of biologists, computer scientists, physicists, and engineers, a picture has begun to emerge of the way in which the nervous system regulates movement.

The human body is capable of an extraordinary range of movements. Years of practice shape the complex skills of professional dancers, pianists, and tennis players. But to neuroscientists, even the simplest everyday movements—reaching for a cup, buttoning a jacket, descending a flight of stairs—present a challenge to scientific explanation. We still do not fully understand how the brain controls these actions, nor can the most sophisticated robotics expert create a machine capable of matching the everyday competence of the central nervous system of the bird, the frog, or the cat, much less that of the human being.

The goal of this essay is to explain what neuroscience has established so far about how the central nervous system (CNS) deals with the complex dynamics of our limbs as they interact with a variable and often unpredictable environment. We will review how scientists have approached the study of movement, the problems they have encountered, and the solutions they have proposed.

Emilio Bizzi is E. McDermott Professor in the Brain Sciences and Human Behavior at the Massachusetts Institute of Technology.

Ferdinando A. Mussa-Ivaldi is associate professor of physiology at Northwestern University.

217

One issue of particular interest to researchers has been the question of how the brain handles the staggering number of mechanical variables involved in even the simplest movement. To illustrate the complexity of this basic problem, consider the analogy of a marionette—a rough imitation of the human body with a head, a trunk, two arms, two hands, two legs, and two feet. Rather than pulling on wires, a modern-day puppeteer uses a computerized control board with a switch connected to each of the marionette's thirteen joints. Each switch can take one of five positions: two for the extreme angles and three for the intermediate values.

To bring the marionette to life, the puppeteer faces the daunting task of mastering and controlling over 5 to the 13th different positions, or approximately one billion. If we now make this simple marionette more like the infinitely more complicated human body—say, by adding ball joints with two angles at the hip, shoulders, hands, and feet—the number of possible positions rises to 5 to the 19th power, or more than ten thousand billion. This analogy gives some sense of the monumental problem handled routinely by the CNS in the ongoing course of motor control.

To further complicate things, there are countless different ways for the CNS to achieve any given goal involving movement. When a reader turns a page, for example, there are a variety of different trajectories the hand could follow, with many combinations of motions at the shoulder, elbow, and wrist. In addition, the single motion of a joint can be "scripted" with numerous patterns of muscle activations. This characteristic of the biological system is called "kinematic redundancy." It means that there is no single solution to a given problem of motor control. How the CNS decides which plan of action to pursue is a difficult and fascinating question for researchers.

Finally, there is the issue of motor learning. In the course of a lifetime, a human being masters a huge repertoire of movements, the memory of which must somehow be stored in the CNS, despite the very real constraints presented by brain anatomy. Even if one were to assume that each of the billions of neurons in the human brain were to represent a posture in the body's repertoire, storage capacity would fall far short of what is needed. How, then, do our brains meet this challenge?

This essay will address the above questions by reviewing some of the experimental findings made over the last few years. First, we will focus on possible ways in which the CNS may produce the forces necessary to generate movements. In this context, we will also consider the problems presented by kinematic redundancy. Second, we will show how motor memories may be represented, stored, and retrieved through the formation of internal models of limb dynamics. Finally, we will review some of the neurophysiological evidence that suggests that motor learning consists of tuning the activity of a relatively small group of neurons. Each of these groups constitutes a "module," which combines with others to produce a vast repertoire of motor behaviors.

THE FORCES THAT DRIVE OUR LIMBS

In the last eighty years, biologists, engineers, and computer scientists have proposed theories to explain how the CNS may produce the forces necessary to generate movements. Generating movements in biological or robotic systems is computationally complex because of the large number of mechanical degrees of freedom of the body. In this section, we will review two sets of ideas: those derived from the field of biology, and those derived from the field of robotics.

A simple yet very common task for our brain is to generate a trajectory of the forearm, involving a temporal sequence of elbow angles from an initial value to a final one. This movement is produced by muscles that together must apply a net force on the elbow joint. From Newton's equation (force = mass × acceleration), we know that the acceleration of an object is proportional to the applied force. Thus in order to move the forearm, the brain must solve a specific problem in physics—that of determining which force must be applied by the muscles in order to produce movement through the desired sequence of angles. Roboticists have called this an "inverse dynamics problem" to distinguish it from the direct dynamics problem, whose goal is to find the trajectory that would result from the application of a known force.

In solving this inverse dynamics problem for a simple movement of the forearm, the brain faces a complex computational challenge. The net force at the elbow is the sum of the forces exerted by all the muscles around a joint; there is thus a degree of arbitrariness in the choice of each muscle's contribution. This situation reflects the ubiquitous "redundancy" that characterizes the motor system and makes inverse dynamics an "ill-posed" problem.[1]

Of course, in everyday life we deal with more challenging tasks than moving the elbow between two angles. Accordingly, the computational problems that the brain must face are more complex than solving Newton's equation for a single joint. For instance, the inertia of the arm, something that our puppeteer must know in order to program the marionette's motion, depends in complex ways upon the angles of the joints. In addition, for purely physical reasons it so happens that the motion of one joint causes a force to be exerted on the neighboring joints. These are factors that the puppeteer must consider. Furthermore, there is the additional issue of redundancy, which while providing flexibility poses a difficult problem for motor control. If we were to ask you to touch a word on this page, you may do so in an infinite variety of ways, each of which involves a different posture for your shoulder, elbow, and wrist. While you make your choice effortlessly, the presence of multiple solutions and the necessity of selecting one among them poses a significant computational challenge for the brain. How does the central nervous system solve all these problems?

Several possible explanations have emerged from studies in robotics and computational neuroscience. At the beginning of this century, Sir Charles Sherrington proposed feedback as a way for the CNS to control a limb's motion.[2] In a feedback system, sensory signals would provide information to the CNS about the position and velocity of the controlled limb at each point in time. If a subject's goal was to reach a desired position with the arm, a feedback-control system would compare the arm's current position with the one desired. The difference between the two positions would serve as a measure of the error at any given time. Once the error was computed, all the brain would need to do is to produce a force directed toward the desired position with an amplitude proportional to the amplitude of the error. This theory of control had the appeal of simplicity.

Sherrington observed that when a muscle is extended, the stretch is countered by an increase in muscle activation. This "stretch reflex" is caused by sensory activity that originates in the muscle spindles—receptors embedded within the muscle fibers. Muscle spindles are well suited for feedback control because they provide direct information on a muscle's length to the CNS. Sherrington hypothesized that voluntary movements were accomplished by combining stretch reflexes with other reflexes in a continuous chain. The theory proposed that movement patterns as complex as walking could be generated by local reflexes without central supervision.

The idea that all movements can be set up by the brain as a chain of reflexes was later found to be simplistic and incompatible with experimental results. If movements were pure reflexes, then we would be paralyzed in the absence of feedback information. In fact, we now know that monkeys and humans can execute various limb movements even after the complete surgical interruption of the pathways that convey sensory information from the limb to the nervous system.

Once it became clear that experimental facts did not support the idea that reflexes alone generated movements, investigators began to search for more effective explanatory alternatives. In recent years, an important contribution to research in biological motor control originated in the field of robotics.

An alternative to the notion of feedback control would be to assume that the CNS explicitly solves the inverse dynamics problem. In other words, the brain computes the forces that the muscles must generate in order to move a limb along the desired trajectory. In theory, this dynamic problem can be addressed only after the trajectory of the joint angles has been derived from the trajectory of the endpoint—that is, after an inverse kinematics problem has been solved. Investigations into robot-control done in the late 1970s and early 1980s have shown that both the inverse kinematics and inverse dynamics problem may be efficiently solved on a digital computer for many robot geometries. On the basis of these studies, John Hollerbach and Tamar Flash put forward the hypothesis that the brain may also be carrying out inverse kinematic and dynamic computations when the arm moves in a purposeful way.[3]

Hollerbach's work in robotics was aimed at finding efficient algorithms for calculating the inverse dynamics of artificial arms.[4] His algorithms are well-organized sequences of elementary operations—additions and multiplications—that lead from the desired trajectories of the limb to the needed forces.

A simpler way to compute inverse dynamics was proposed by Marc Raibert in 1977.[5] Raibert started from the observation that the inverse dynamics problem can be represented as the operation of a memory that associates a set of forces with each specific state of motion of the arm. In his approach, the values of the various torques for each possible value of position, velocity, and acceleration of the limb are stored in a computational device that computer scientists call a "look-up table." Unfortunately, the huge demand for memory size makes the look-up table an impractical solution in the biological context.

The work of Raibert and Hollerbach had the merit of showing that the inverse dynamics of limbs may be computed for the robot with a reasonable number of operations and with reasonable memory requirements. However, this work provided no direct evidence that the brain engages in such computation. Furthermore, on a purely theoretical level, explanations based on computing inverse dynamics are unsatisfactory because there is no allowance for the inevitable mechanical uncertainty associated with a limb's interaction with the environment. Living organisms, unlike conventional robots and computers, generally do not operate on the basis of some predefined program. Instead, they learn from experience. As a result, the theories from early robotics, which focused on how a system can be programmed to compute dynamics, did not shed much light on how the brain could learn to deal with the dynamics of limbs operating in the context of a dynamically changing environment.

MOTOR LEARNING: THE ROLE OF INTERNAL MODELS

The focus on learning from experience as a means of acquiring motor skills has gained great strength in recent years. This new approach derives in large part from theoretical and experimental studies on networks of idealized neurons. A number of theoretical studies have shown that when networks of artificial neurons are exposed to repeated motor commands paired with their sensory consequences, learning of fairly complex motor tasks may take place without the need for explicit programming. The learning results from a change in the internal structure of the artificial network, specifically a change in the connectivity among its elements.

On the basis of these results, scientists have proposed that similar processes might be present in the central nervous system. The hypothesis is that learning is the result of repeated exposures to sensory signals coming from the moving limbs as they interact with the environment. The repeated sensory signals are funneled to the motor areas of the central nervous system, where signals that activate the muscles are produced. The actions produced by the activity of the motor areas are initially imprecise, but a feedback mechanism produces a gradual convergence on the correct solution. Ultimately, this iterative process would lead to the establishment of an internal representation of the task through the gradual change in the synaptic strength of the neurons of the motor areas. If the task is that of moving a limb, for example, the outcome of learning would be the formation of an internal model of the limb's dynamics. The internal model, according to this view, is embedded in the newly formed connectivity of a group of neurons. The activity of this group of neurons generates the neural impulses necessary for the execution of the learned motor task.

The experimental results obtained by Reza Shadmehr and Ferdinando Mussa-Ivaldi support the notion of internal models.[6] Their experimental setup was simple: human subjects were asked to make reaching movements in the presence of externally imposed forces. These forces were produced by a robot whose free endpoint was held as a pointer by the subjects. The subjects were asked to move the pointer toward a number of visual targets. Since the forces produced by the robot significantly changed the dynamics of the reaching movements, the subjects' movements were at first grossly distorted when compared to the undisturbed movements. However, with practice, the subjects' hand trajectories in the force field converged to a path similar to that produced in the absence of any force field.

In other words, the subjects learned to compensate for the applied forces. In order to investigate the neural changes underlying this type of motor learning, Shadmehr and Mussa-Ivaldi devised a simple but revealing experimental manipulation. After the subjects had learned to compensate, the researchers removed the perturbing force for the duration of a single movement. The resulting trajectories, named "aftereffects," were approximate mirror images of the distorted movements that were observed when the subjects were initially exposed to the forces.

The emergence of these aftereffects suggests that the central nervous system composes an internal model of the external force field, a model that generates patterns of compensating forces that anticipate the forces that had perturbed the moving hand.

It is of interest to ask what the properties of the internal model might be, and whether the model could generalize to regions of the work space where the perturbing forces had not been experienced. Recent experiments by Francesca Gandolfo and colleagues were designed to test whether motor adaptation generalized to regions of the work space where no training had occurred.[7] In these experiments, subjects were asked to execute point-to-point planar hand movements between targets placed in one section of the work space. The subject's hand grasped the handle of a robot, which was used both to record and disturb their trajectories. Again, as in the experiments of Shadmehr and Mussa-Ivaldi, the adaptation was quantified by the degree of aftereffect observed when the perturbing forces were discontinued.

Gandolfo found that aftereffects were present, as expected, along the directions where subjects had been trained, but the magnitude of the aftereffects diminished smoothly with increasing distance from the trained locations. This finding indicates that the ability of the CNS to compensate for external forces is restricted to those spatial locations where perturbations have been experienced by the moving arm.

In summary, the work of Shadmehr and Mussa-Ivaldi and of Gandolfo and his collaborators has shown that subjects adapt to a new environment by forming a representation of the external force field that they encounter when making reaching movements. Does this representation form an imprint in long-term memory? Recently Thomas Brashers-Krug and his coworkers investigated this question by exposing their subjects' movements to forces that interfered with the execution of reaching to a target.[8] After some practice, these subjects were able to guide the cursor accurately to the targets despite the interfering forces.

Twenty-four hours after learning the task, one group of subjects was tested with the same disturbing forces and demonstrated not only retention of the acquired motor skill but also additional learning. Surprisingly, they performed at a significantly higher level the second day than they had the first.

A second group of subjects was trained on day one, like the first group, to execute reaching movements with a perturbing field (task A). Immediately afterwards, on the same day, these subjects were trained to execute the same movements with perturbing forces in the opposite direction (task B). When these subjects were tested on a subsequent day, Brashers-Krug's team found that retention of task A had been significantly impaired by exposure to task B. This phenomenon is known as "retrograde interference." In a later experiment, the same researchers found that retrograde interference decreased monotonically with time as the interval between task A and B increased. When four hours passed before task B was learned, the skill learned in task A was completely retained; apparently, the initial learning had consolidated. What is remarkable in these results is that motor memory was transformed, with the passage of time and in the absence of further practice, from an initial fragile state to a more solid state.

Taken together, the experiments just described indicate several things: 1) There was a certain degree of specificity in the learning of a simple motor task. The internal model that the subjects learned was restricted to that part of the space where interference had been experienced. The same external forces could not be handled in a different part of the work space. 2) There was an enhancement of the learned task that did not depend upon practice, but only on the passage of time. 3) There was a process of consolidation of learning that took four hours at a minimum. The consolidation was not dependent upon practice; it was an internally generated event.

We conceive of the internal model as a newly formed rearrangement of synaptic contacts among a group of neurons. It is theoretically possible that a given neuron may participate in a number of different groups, each supporting different internal models. Given the large number of synapses on the surface of neurons, this sharing could sustain a large number of internal models. As an alternative, the internal models could be conceived not as independent monads but as entities that can be combined into bigger assemblies when more demanding motor tasks are faced by the body. The study of the brain circuitry at the cellular level will undoubtedly provide new evidence on these issues in the near future.

THE NEURAL SUBSTRATE OF INTERNAL MODELS

In the previous section, we outlined the concept of an internal model for the dynamics of a moving limb. Now, we will describe the physiological evidence supporting the theory that the brain areas responsible for generating motor commands also serve as the sites for the storage and retrieval of motor memory. This linkage is consistent with the view that the brain circuit that has learned a task becomes the command center for expressing that task.

There are several examples of the intermingling between control functions and motor memory in the cortex. Brian Benda and his colleagues have reported some of the most direct evidence for the development of new patterns of activity in the cells of the motor area of the frontal lobe, an area named M1.[9] It should be pointed out that M1 is a key motor area, and damage to it profoundly disturbs the ability to produce voluntary movements. Benda's most striking result showed the gradual appearance of activity in cortical neurons of the M1 area in monkeys practicing arm movements against disturbing forces. These neurons displayed activity related to the production of forces that compensated for externally imposed interference. Remarkably, the same neurons were inactive before the application of the disturbance, but they remained active after the disturbance was removed. This effect is consistent with the hypothesis that neurons in M1 operate as memory elements. Similar results have been reported by Steven Wise, who used the technique of single-cell recordings but with different behavioral paradigms.[10]

John Martin and Claude Ghez reached similar conclusions with the use of a pharmacological ablation of M1.[11] They demonstrated that after the inactivation of this area, their experimental animals could not learn to correct the trajectories of perturbed limbs. In recent years, investigators have demonstrated learning in M1 with imaging techniques, namely positron emission tomography and functional magnetic resonance.[12]

In addition to the primary motor cortex (M1), other cortical areas of the frontal lobe, namely the premotor cortex and the supplementary motor areas, have been found to be involved in motor learning, either in conjunction with M1 or in isolation. In particular, the premotor cortex specializes in the learning and retention of visuo-motor tasks.[13] The supplementary motor areas seem predominantly concerned with sequence learning and conditional learning.[14]

Imaging studies have indicated that the prefrontal areas are also involved in motor learning. In particular, the experiments of Shadmehr and Henry Holcomb used the Brashers-Krug paradigm to gather evidence suggesting that the formation of an internal model of a perturbing force is associated with increased activity in the prefrontal cortex.[15] However, towards the completion of the learning task, recall of the learned internal model became correlated with increased blood flow in other cortical and subcortical areas, such as the premotor cortex and the cerebellum, and with decreased blood flow in the prefrontal cortex. A possible interpretation of this shift in blood flow is that the prefrontal cortex is a temporary storage area for sensory-motor associations.[16]

Taken together, these studies show that the motor cortical areas are linked to processes involved in motor learning—a result that implies that the circuitry of these areas may have the capacity to reorganize its functional properties. This capacity to reorganize depends upon the formation of new synapses. William Greenough reported that the dendritic branches of cortical neurons in M1 increase in number with motor training.[17] Presumably new synapses are formed on these branches. This possibility was confirmed by the recent report from Asanuma's group that electrical stimulation of the thalamus increases the density of synapses in the motor cortex.

MODULAR ORGANIZATION OF THE MOTOR SYSTEM

The evidence discussed in the previous sections suggests that the central nervous system is capable of representing the dynamic properties of limbs as well as the environment with which our limbs interact. Presumably, a representation is built upon some elementary building block, or "module," in the same way that sentences are composed of words. How is the representation accomplished? Recent electrophysiological studies of the spinal cord by Emilio Bizzi and his coworkers suggest how the CNS transforms the internal model into action.

The spinal cord is the final output stage of the motor system. Every muscle is innervated by motoneurons located in the ventral portion of the spinal gray matter. We may regard this system of motoneurons as comparable to the switchboard that drives a marionette. But there is more than this switchboard in the spinal cord. In addition to the motoneurons, the spinal gray matter contains a large population of nerve cells, called interneurons, whose functions are not yet fully understood. We know that these interneurons are capable of forming connections with motoneurons that innervate several different muscles.

In experiments performed by Bizzi, Mussa-Ivaldi, and Simon Giszter, the activity induced by the electrical stimulation of the spinal interneurons of the frog was found to spread to several groups of motoneurons.[18] This distribution of activity was not random but imposed a specific balance of muscle contractions. The mechanical outcome of the evoked synergistic contraction of multiple muscles was captured by a *force field*; the activation of a group of muscles generated a force that was recorded by a sensor at the endpoint of the limb. This force vector changed in amplitude and direction according to the position of the limb. Following stimulation of the spinal cord, the resulting force field converged toward a location in the reachable space of the limb— a stable equilibrium point. At this location, the force vanished and a small displacement of the endpoint in any direction caused a restoring force to appear. Thus this location acted as an attraction point for the limb in the same way as the bottom of a teacup is an attraction point for a marble rolling inside the cup. The analysis of the force field induced by the stimulation of the spinal interneurons revealed that such activation leads to a stable posture of the limb.

After the force field was identified, the stimulating electrodes were placed in different loci of the lumbar spinal cord, which activated a number of groups of leg muscles. After mapping most of the premotor regions in the lumbar cord, Bizzi, Giszter, and Mussa-Ivaldi reached the conclusion that there were at least four areas from which distinct types of convergent force fields were elicited.

Perhaps the most interesting aspect of the investigation of the spinal cord in frogs and rats was the discovery that the fields induced by the focal activation of the cord follow a principle of vectorial summation. When two separate sites in the spinal cord were simultaneously active, the resulting force was the sum of the forces induced by the separate activation of each site. This discovery led to a novel hypothesis for explaining movement and posture based on combining a few basic elements. The few force fields stored in the spinal cord may be viewed as representing motor primitives from which, through superposition, a vast number of movements can be formed by impulses conveyed by supraspinal pathways. According to this view, the supraspinal signals would establish the level of activation for each module. By means of mathematical modeling, Mussa-Ivaldi and Giszter along with subsequent work by Alexander Lukashin verified that this view of the generation of movement and posture is capable of accounting for a wide repertoire of motor behaviors.[19]

These experiments suggest that the circuitry in the spinal cord—and perhaps also in other areas of the nervous system—is organized in independent units, or modules. While each module generates a specific field, more complex behaviors may be produced by superposition of the fields generated by concurrently active modules. We may therefore regard these force fields as independent elements of an internal model of dynamics. Recent simulation studies by Mussa-Ivaldi have demonstrated that by using this modular representation—that is, by adding convergent force fields—the central nervous system may learn to reproduce and control the dynamics of multijoint limbs in the context of a dynamic environment.

CONCLUSION

In this review, we have explained how the brain deals with the complex dynamics of our limbs as they interact with a variable and often unpredictable environment. We have shown that it is computationally difficult to produce the forces that drive our limbs because of the staggering number of mechanical variables involved in even the simplest movement. The kinematic redundancy of our motor system means that there are many different ways for the central nervous system to achieve an intended motor goal. While providing flexibility of motion, the redundancy creates for the CNS the difficult problem of deciding which plan of action to pursue.

We have furthermore proposed a theory based on internal models to explain how the CNS controls limb dynamics. Through repeated exposure to sensory signals coming from the moving limb during the acquisition of a motor task, there is a gradual change in the synaptic strength of the neurons of the motor areas. The outcome of this process is the formation of an internal model of limb dynamics. We have presented experimental evidence demonstrating that the formation of internal models as a means of acquiring motor skills is a more plausible hypothesis than those proposed in the past.

Finally, we have stressed the modular organization of the motor areas of the CNS. At the cortical level, we have demonstrated the tuning of small groups of neurons during motor learning; at the spinal cord level, our work has shown the existence of modules, which can be combined to produce different motor behaviors.

ENDNOTES

[1]Problems that are encountered in physics may be classified into well-posed and ill-posed problems. A well-posed problem is one for which there exists a unique solution that depends continuously upon the data. By contrast, ill-

posed problems may have either no exact solution or a multiplicity of solutions, and these solutions may change abruptly for certain values of the data.

[2]Charles Sherrington, "Flexion-reflex of the Limb, Crossed Extension Reflex and Reflex Stepping and Standing," *Journal of Physiology* 40 (1910): 28–121.

[3]John Hollerbach and Tamar Flash, "Dynamic Interactions between Limb Segments during Planar Arm Movements," *Biological Cybernetics* 44 (1982): 67–77.

[4]John Hollerbach, "A Recursive Formulation of Lagrangian Manipulator Dynamics," *IEEE Transactions on Systems, Man, and Cybernetics* SMC-10 (11) (1980): 730–736.

[5]Marc Raibert, "Analytical Equations versus Table Look-up for Manipulation: A Unifying Concept," in proceedings of the IEEE Conference on Decision and Control, New Orleans, 1977.

[6]Reza Shadmehr and Ferdinando Mussa-Ivaldi, "Adaptive Representation of Dynamics during Learning of a Motor Task," *Journal of Neuroscience* 14 (1994): 3208–3224.

[7]Francesca Gandolfo, Ferdinando Mussa-Ivaldi, and Emilio Bizzi, "Motor Learning by Field Approximation," *Proceedings of the National Academy of Sciences* 93 (1996): 3843–3846.

[8]Thomas Brashers-Krug, Reza Shadmehr, and Emilio Bizzi, "Consolidation in Human Motor Memory," *Nature* 382 (1996): 252–255.

[9]Brian Benda, Francesca Gandolfo, Chiang-Shan Li, Matthew Tresch, Daniel DiLorenzo, and Emilio Bizzi, "Neuronal Activities in M1 of a Macaque Monkey during Reaching Movements in a Viscous Force Field," *27th Annual Society of Neuroscience Abstracts* 23 (1997): 607.12.

[10]Steven Wise, S. Moody, K. Blomstrom, and A. Mist, "Changes in Motor Cortical Activity during Visuomotor Adaptation," *Experimental Brain Research* (forthcoming).

[11]John Martin and Claude Ghez, "Task-related Coding of Stimulus and Response in Cat Red Muscles," *Experimental Brain Research* 85 (1991): 373–388.

[12]S. Grafton, J. Mazziotta, S. Presty, K. Friston, R. Frackowiak, and M. Phelps, "Functional Anatomy of Human Procedural Learning Determined with Regional Cerebral Blood Flow and PET," *Journal of Neuroscience* 12 (7) (1992): 2542–2548. See also A. Karni, G. Meyer, P. Jezzard, M. Adams, R. Turner, and L. Ungerleider, "Functional MRI Evidence for Adult Motorcortex Plasticity during Motor Skill Learning," *Nature* 377 (1995): 155–158.

[13]Richard Passingham and Ulrich Halsband, "Premotor Cortex and the Conditions for Movement in Monkeys (Macaca fascicularis)," *Behavioral Brain Research* 18 (1985): 269–277. See also Michael Petrides, "Motor Conditional Associative-learning after Selective Prefrontal Lesions in the Monkey," *Behavioral Brain Research* 5 (1985): 407–413.

[14]Hajime Mushiake, Masahiko Inase, and Jun Tanji, "Neuronal Activity in the Primate Premotor, Supplementary, and Precentral Motor Cortex during Visually Guided and Internally Determined Sequential Movements," *Journal of Neurophysiology* 66 (1991): 705–718.

[15]Reza Shadmehr and Henry Holcomb, "Changing Activations in the Prefrontal Cortex during the Time-dependent Phases of Motor Memory Formation," *Journal of Neuroscience* (forthcoming).

[16]Joaquin M. Fuster, *The Prefrontal Cortex: Anatomy, Physiology, and Neuropsychology of the Frontal Lobe* (New York: Raven Press, 1989). See also Patricia S. Goldman-Rakic, "Topography of Cognition: Parallel Distributed Networks in Primate Association Cortex," *Annual Review of Neuroscience* 11 (1988): 137–156.

[17]William Greenough, "Structural Correlates of Information Storage in the Mammalian Brain: A Review and Hypothesis," *Trends in Neuroscience* 7 (1984): 229.

[18]Emilio Bizzi, Ferdinando Mussa-Ivaldi, and Simon Giszter, "Computations Underlying the Execution of Movement: A Biological Perspective," *Science* 253 (1991): 287–291.

[19]Ferdinando Mussa-Ivaldi and Simon Giszter, "Vector Field Approximation: A Computational Paradigm for Motor Control and Learning," *Biological Cybernetics* 67 (1992): 491–500; Alexander V. Lukashin, Bagrat R. Amirikian, and Apostolos Georgopoulos, "Neural Computations Underlying the Exertion of Force: A Model," *Biological Cybernetics* 74 (1996): 469–487.

Marcel Kinsbourne

Unity and Diversity in the Human Brain: Evidence from Injury

The laws of nature are constructed in such a way
as to make the universe as interesting as possible.
                                    —Dyson's Principle

D OES DYSON'S PRINCIPLE APPLY to the universe within? If the human brain were homogenous, with all parts equally involved in all intelligent activities, then its manner of operation would be inscrutable and not very interesting. (Such a suggestion has been made under the heading "mass action," but it has been rejected—see Vernon Mountcastle's essay in this volume.) If it were instead composed of discrete components (or "modules") like a machine, then in principle its composition would not be hard to determine, though how such a conglomerate might have evolved would be perplexing.[1] In reality the brain is more interesting than either of those blueprints suggests. It presents a paradox: It is a completely connected nerve net, and yet it is highly differentiated in its parts. How can those two characterizations be reconciled?

COGNITIVE NEUROPSYCHOLOGY

Nobody knows how the brain works. But even a staggeringly complex system can be implemented by combining a limited set

---

Marcel Kinsbourne is Professor of Psychology at New School University.

of building blocks, and evolution notoriously relies on a limited inventory of tactics, variously applied. We are familiar with some of the design characteristics of the neural substrate of human cognition, largely through one hundred and fifty years of observing the effects of brain injuries on people's ability to experience, think, and act. Supplemented by the patients' subjective reports, these controlled observations indicate fault lines along which cognition falls apart, thereby offering hints as to what its components, or primitives, might be. This branch of neuroscience is called cognitive neuropsychology. Additionally, many of these case studies contribute to a mapping of cognitive functions on the brain via evidence about the localization of the causative lesions (brain-behavior relationships). The process involves zeroing in on the cognitive operation–brain localization pairing: The patient with a focal injury fails at certain tasks; the investigator hypothesizes about the aspect of the tasks that presented difficulty; he then designs novel tests that more selectively probe for the hypothesized mental operation; and finally he determines whether a more reliable behavior-brain relationship is understood than before. Step by step, the nature of the cognitive primitive to be ascribed to the damaged location is crystallized.

I shall present neuropsychological observations—usually my own, but consistent with those of others—that will demonstrate that the human forebrain is exquisitely functionally differentiated and yet quite integrated. Next, I shall show that the way functions are organized in the brain corresponds in principle to how the cerebral nerve net is organized neuroanatomically. I shall then offer some suggestions as to how this organ, which is both integrated and differentiated, might work.

Most informative are local and specific injuries. Examples include limited areas of the brain deprived of circulating blood on account of the occlusion of an artery (stroke), areas replaced or compressed by a tumor, and areas that disintegrated under the impact of a blow or a missile or in the wake of a localized infection (brain abscess). Injured in such ways, people may no longer be able to do some things that they could do previously. They now exhibit a selective cognitive deficit, whose nature depends more on where in the brain the injury is located than what caused it. The smaller the lesion, the more likely it is to compromise only a very few of the things that a person can do, leaving the rest of the cognitive profile roughly intact. In contrast, large lesions are apt to impair a whole cognitive domain, such as language. Small lesions could compromise a subskill within that domain, leaving the other subskills functional (for instance, the ability to name objects aloud or, in another lesion location, to repeat sentences). My purpose is to illustrate what types of specializations exist, how they relate to each other at the cognitive level, and how in general they cluster in the forebrain.

FINE-GRAINED DISSOCIATIONS

Highly selective deficits have been documented in every aspect of cognition. Consider a patient with damage to the back of the left cerebral cortex, who can write but cannot read what he has written (alexia without agraphia).[2] Or a patient with a more extensive posterior lesion, involving both sides, who cannot recognize common objects when they are shown to him yet can identify them by touch, copy them, and draw such objects from memory without a model (associative object agnosia).[3] These are "dissociations" within a single person. This sort of patient performs normally on all but one type of task. One can also compare people with different brain injuries. When one brain-injured person fails on task A (say, spelling aloud) while succeeding on task B (say, spelling in writing), whereas another patient performs in just the reverse way, a "double dissociation" between the two activities is said to obtain.[4] The same applies to the patients who can understand what is said but not repeat it (conduction aphasics) and those who can repeat speech verbatim but have no idea what it means (transcortical motor aphasics).[5] Evidently the brain handles each one of the two dissociated activities separately.

The fine grain of cerebral specialization is remarkable. There are patients who can recognize faces but not letters (alexia), or letters but not faces (agnosia for faces). One patient's difficulty was limited not just to handwriting (apraxic agraphia), but solely to writing in cursive.[6] A patient could identify objects, explain their use, but not retrieve their names; he could, however, name them when he felt them in his hand.[7] Another patient could not name the colors or recognize them when named, yet he could recognize and name black, white, and gray (color agnosia).[8] When competent investigators report a dissociation in a single case, that counts as an existence proof—an aspect of brain functioning has been revealed. Brain theory must accommodate the finding.

COARSE-GRAINED DISSOCIATIONS

Right-Left

The instances that I have cited are notable for their discreteness. The deficit scales up with larger lesions. To study these, group comparisons are more often used than single-case experimental designs, so one may compare right-lesioned patients with those whose injuries are on the left. Splinter deficits that result from fine-grained dissociations will be submerged in the statistical treatment of the group data, but broad distinctions emerge. We learn that most people's left cerebral hemisphere is concerned with language, their right with spatial orientation. More broadly still, the left caters to sequential analysis and the generating of action sequences, the right to setting such activities into a spatial framework.[9] Most general of all, the left hemisphere controls motivated approach sequences(handling, eating, and so on), progressively focusing and acting upon the target; the right hemisphere is more involved with the person's movement through the intervening space and the spatial background of the target. The left hemisphere's activities can be context free, whereas the activities of the right are context bound. The hemispheres are complementary in their functioning.

Unsurprisingly, the left and right hemisphere mediate different types of attention, roughly focal and global, respectively.[10] Patients with left-hemisphere lesions, who therefore rely largely on the right hemisphere, can perceive and reconstruct the general shape of a thing, but not in analytic detail. Patients with right-hemisphere lesions can serially extract the details, but not articulate them into a coherent whole.[11] Global attention encompasses the whole display and is bound to that context. Focal attention disregards context while it targets specifics.

Comparable observations have been made at the level of concepts. Patients whose right hemisphere was temporarily inactivated (soon after electroconvulsive shock treatment for depression) could accept and respond to questions (syllogisms) that assume counterfactuals, such as "Canada is on the equator. Is it hot in Canada?" They could disregard what they knew to be the case and assume the hypothetical. When the same patients' left hemispheres were inactivated, they protested the factual misstatement and refused to respond. They were context-bound to their knowledge base and could not decontextualize even conceptually. (What happens in the intact individual, in whom both hemispheres are available for use? A compromise. The patients responded more consistently to the counterfactual syllogisms when operating only on their left hemispheres than when operating on both.)[12]

The above conclusions are drawn from studies of patients with damage varying in extent but falling short of implicating the whole hemisphere. Total removal of a hemisphere is rare. But it is possible and sometimes clinically appropriate to simulate hemispherectomy for a short time, measured in minutes. One technique is one-sided electroconvulsive shock (ECS). This can relieve depression that resists medications. Another is the injection of intracarotid amobarbital (ICA). The hemisphere on the injected side (save the rearmost portion) is inactivated for about three minutes. The dysfunction that results affords a preview of what would occur if surgery is performed on that hemisphere. This procedure is most commonly used to screen for possible adverse effects on memory from a temporal lobectomy for intractable epilepsy.

The use of ECS and ICA reveals that people remain conscious when only one hemisphere is active, regardless of which one it is. Therefore, an indispensable "consciousness module" cannot exist in either hemisphere. Cerebral lesions in various locations restrict the richness of the contents of consciousness, but no one focal lesion can abolish it outright. Consciousness appears to be a property of the global network, not of an elite localized conscious awareness system. With respect to language, we learn that one in five left-handers programs language in the right hemisphere, and just as many use both.[13] What is more, ICA performed on patients rendered aphasic by left-hemisphere strokes reveals that in some cases the right hemisphere takes over the language role.[14] The language function must have arisen from variations on the properties of the global network. Right-hemisphere compensation would not be possible if it relied on unique, recently evolved circuitry. Correspondingly, we know from anatomical studies that the language cortex relies on evolutionarily ancient cortical circuitry.[15]

Front-Back

An equally sweeping contrast obtains between the front and the rear (anterior and posterior) cortical regions. The prefrontal cortex exerts control over posterior activities by means of corticocortical connections that transmit along separate parallel lines in both directions.[16] It thereby brings influences to bear on the local posterior cortex—from the physical and social environment, and from the person's wishes, beliefs, preferences, and apprehensions (arising in the limbic cortex).[17] Correspondingly, rather than impairing a specific skill, prefrontal lesions handicap mental agility, for example, the ability to change one's mind-set rather than persevere with an unsuccessful strategy. In this way the prefrontal cortex, and specifically its dorsal subdivision, enables the setting of priorities depending on the situation.[18] In contrast, orbitofrontal lesions release impulsive responding, the uncensored drive-directed activity of the remaining intact brain. The lateral prefrontal cortex does cost-benefit accounting of the proposed act while the individual hesitates. Absent this accounting, responses are quick, incorrect, or hazardous. Phineas Gage, the famous prefrontally injured railroad worker, is a classical example. Premorbidly a conscientious worker, after an iron bolt tore through both his prefrontal cortices he became feckless and disorganized, unreliable and unemployable. His intelligence remained intact.[19]

Yet the prefrontal cortex is not a general "central executive." It is comprised of a set of control systems, each with its separate target.[20] As their connectivities indicate, different prefrontal areas control different activities that are separately localized posteriorly. In each case, the prefrontal contribution enables the individual to overcome primitive, preprogrammed responding when that would be maladaptive. It also applies in the temporal domain to action sequences. For the brain, unlike for a machine, ceasing to do something is a positive act, in the same way that starting it is. Processing sequences do not just run dry in the sands, and some patients exhibit run-on behavior. An example is the jargon aphasic, who rattles on well after he has, at least to his own satisfaction, responded to the question. He may even put a stop to his harangue by saying "stop" (Luria's verbal control of behavior).[21] In broad outline, left prefrontal lesions impair planning and right prefrontal lesions impair interpersonal relating (leading to so-called pseudopsychopathy). Both describe Phineas Gage. Again we see an overall network—this time prefrontal, with a particular potential for steering cognition—that is partitioned into subnets that exert the same general type of influence in different contexts and toward different goals.

GLOBAL IMPAIRMENTS

Intact people are able to do more than one thing at a time; how well they do so depends on whether both activities demand attention (i.e., are not automatized). If they both demand attention, then whether they can be performed in parallel depends on what I have called the "functional cerebral distance" between them.[22] The less neurally interconnected the two areas that guide the activities in question are (i.e., the further apart they are), the more easily the activities can be performed at the same time. Activities that are based on highly interconnected areas interfere with each other to the point that a person can only do one at a time. This illustrates the fact that how the brain is organized constrains what people can do. Vernon Mountcastle's essay in this issue refers to the ongoing study of functional space at the neural level.

If the global network is extensively depleted of neurons, as in the dementias and to a lesser extent even in so-called normal aging, regressions occur that are domain-general.[23] Modular discontinuities are not respected. The demented person increasingly relies on familiar, well-rehearsed, or biologically prepotent routines. The brain states that remain available exhibit inertia. Once implemented, a brain state is unduly persistent, which causes the person to exhibit a rigid and inflexible personality. He relies on well-trodden paths of attitude and action; in effect, he becomes like himself, only more so. The inertia of state transitions plays out not only as a difficulty in adapting a mental set to changing conditions, but even as a general slowness in processing information. At its extreme, dementia restores constraints that bind the infant, as when primitive reflexes of infancy resurface—movement patterns (synergisms) that normal adults automatically suppress. The restriction in cognitive state space is compounded by restriction in movement state space.

NESTED COGNITIONS, NESTED NETWORKS

We emerge with an understanding of cerebral organization as an architecture of nested functional units. Within an overall sequential analytic mode, the left hemisphere generates word sequences in one of its parts, action sequences in another, and sequential identification (for instance, of letters) in yet another. Within subareas of the respective areas, it enables both recognition and expression. In subareas of those subareas, it enables material-specific processing, for example, of action words, color names, names of animals, or the connotations of "right" and "left." Within its overall spatial relational mode, the right hemisphere enables orientation in ambient space in one of its parts and pattern perception in another. In subareas of these subareas, it permits subskills like map reading, face recognition, or identifying sketched shapes ("perceptual closure"). A recursive organization emerges; superordinate processing modes differentiate into distinctive domains of functioning, and these again into different specific applications. I envisage a relatively undifferentiated whole that progressively differentiates (in child development? in evolution?) into the rich set of specifically human potential skills and alternative strategies.

The recursive organization of the "cognitive profile" finds its counterpart in the recursive organization of the cortical neuronal network. The cortex is conceptualized as a recursive network or "net of nets," constructs that are vividly realized in cerebral organization: The global network is composed of separate parallel cortical "trends" (and patches of multimodal cortex).[24] Trends are sequences of processing stages. One instance is the "dorsal visual stream" that interfaces with visual input, especially from the peripheral fringes of the visual field in visual area one (V1). V1 is the first cortical relay for visual input; its more internal anchor is the less differentiated and earlier-evolved parahippocampal gyrus. Another trend, the ventral stream, also relates V1 input but more from the focal center of the visual field to less differentiated and earlier-evolved stages, culminating in inferotemporal cortex. Anatomical minutiae apart, the implication for brain organization is that trends are composed of individual sequentially connected processing units called stages (with each stage consisting of three parallel subsections). The latter in turn consist of sets of columns of cortical cells. So I (that is, the brain that is me) am a network (global) of networks (trends) of networks (stages) of networks (columns). The recursive organization of cognition arises naturally from the recursive organization of cortical nets.

The term "stream" is misleading in that it implies a unidirectionality of flow, which is not the case. Trend is a more neutral term that can accommodate the fact that activation can flow in both directions: from the outside in, signaling a perturbation of the system, and from the inside out, preforming the anticipated sensory consequence of the act in progress.[25] The two waves of activation meet, forming a standing wave across the elements of the trend. The conscious content of the activation pattern that results is an inextricable amalgam of represented anticipation and represented perturbation.

It used to be believed that much of the human cerebrum was dedicated to higher mental functions, and only a small part to sensorimotor input and output. We now know that all of the cortex originally evolved as a series of adaptations to the requirements for sensation and action. Our ability to manipulate and abstract is superimposed on this wealth of parallel sensorimotor representation, not localized in a separate part of the brain. Again, evolution pours new wine into old bottles.

REALITY CHECK

Are the localizations reliable? Can one directly interpret cognitive deficits in terms of the preexisting skills that the affected areas of the brain normally contribute? In short, are the lesion effects "transparent" with respect to the underlying mechanisms? Given the vagaries of localization, of plasticity, of compensatory activity, of premorbid strengths and weaknesses, can they be?

Lesion site and type of deficit do not exactly correspond in different patients. But neither does a given "splinter deficit" result from lesions in widely different locations in different patients. We are not mistaking the electric cord for the light bulb—cutting the cord, seeing the light go out, and supposing that the light source was where we cut. If we were, we would observe the same narrow deficit as a result of damage in many different parts of the brain. This does not happen. The fine grain of human cortical specialization is unquestionable, as demonstrated time and again by the controlled study over years and decades of highly selected, focally injured patients.

The brain operates in terms of interactive groups of neurons that excite or inhibit each other, two-way connections between neuronal assemblies, and neurohumors that transmit coded information (neurotransmitters) or adjust the set point for change in the rate at which particular cell assemblies oscillate (neuromodulators). We are far from being able to identify the neurocognitive primitives, the particular patterns of firing for particular cell assemblies that underlie any human activity at all. But we can propose generalizations about how the system works as well as claims about how it does not work. I base the latter proposals on what I call "nonexistence proofs."

NONEXISTENCE PROOFS

The nonexistence proof is an essential corrective to the extravagances of inductive reasoning. By carefully selecting his instances, a theorist can support any one in too wide a range of possible solutions. The nonexistence proof invokes the negative instance. It enables us to weed out solutions that are nonstarters and alerts us to some ways in which the brain might not work.

In neuropsychology the nonexistence proof specifies conceivable deficits that do not in fact occur. But how can we prove a negative? How can we ever be sure that a particular form of malfunction does not, and could not, occur? By all means, we are constantly surprised by new findings, but the following remains a viable working hypothesis: If a particular malfunction has not been discovered after a century and a half of diligent search by neuropsychologists the world over, it probably does not occur. Our preferred model of brain function should then be one in which such malfunctions would indeed not occur.

The nonexistence proofs that I offer point in an interesting direction. Rather than being assembled piecemeal and glued (conjoined, integrated) together, the percept, construct, utterance, or intention gradually differentiates out of the preexisting brain state.[26] Diversity is continually being carved out of the existing unity. The operative question is not "How are the details assembled into a whole?" but rather "How is the whole reshaped to incorporate the details?" What follows are some mistakes patients do not make.

Patients with focal brain lesions do not violate the rules that governed their premorbid performances. They remain guided by preexisting parameters but become less specific, slower, and less stable in their responses. Preexisting constraints become more stringent, and the boundary conditions for the impaired behavior become more limiting. Properly scrutinized, apparently qualitative differences turn out to be quantitative differences. Patients retain high-frequency responses (e.g., names of familiar things) while losing low-frequency ones. Patients continue to operate within the relevant domain even if they mistake or misspecify the exemplar, with errors that conserve the implicated domain being the most common. No real focally damaged patient mistook his wife for a hat. He might, however, mistake her for another person or mistake his hat for his scarf. This point is illustrated by "deep dyslexia," in which words are often misread, not for words that look or sound similar, but for words with similar meanings.[27] In the domain of action, an apraxic person who fails to make a fist does not do handstands instead; he does something of the same general nature as the act that was requested. The patient with a visual-recognition deficit (agnosia) may know that he is viewing an animal, but not which one. He may even distinguish possible from impossible pictured objects, without knowing the identity of the possible objects. In short, in the impaired domain, the patient's cognitions are simplified, primitive, even rudimentary, but not fanciful, perverse, or wildly irrelevant. These principles befit a recursive organization. The totally lesioned area, itself incorporated, is nested within a broader area that deals with the same domain as the lesioned "module," though in a less differentiated manner.

Within a specialized network, individual responses are not further structurally differentiated and segregated; rather, the full set of responses arises from patterned activation of the entire network unit. Thus within the affected domain the deficit applies to the type. Individual tokens do not become selectively unavailable. A patient with word-finding difficulty (anomia) does not selectively have trouble with all words that begin with the letter *f* or are bisyllabic or have an unpleasant connotation, let alone with one particular word, sparing the rest. I conclude that the same network handles all tokens of the implicated type. What constitutes a type, however, is determined by the brain. The special facility for face recognition makes faces a type; within that type, there are no selective losses. The hand in action is perhaps also a type. But apparently no comparable aggregation of circuitry deals with bodies, or autographs, or infants that do not yet walk. There is no general rule by which one could predict what aggregates and what dissociates in the brain.

The principle of domain-specific dedifferentiation is well illustrated in the way in which aphasics generate sentences. Among the conceivable subtypes of lesion-induced language disorder, one is conspicuously absent: a jumbling of the sequence of uttered words. Words may be omitted or simplified, for instance, resulting in lost inflection, but no aphasic "order the himself wrong in expresses." I conclude that one way that sentences may not be formed is by lining up words like beads on a string. Instead, I infer that sentences differentiate out of less-specific preconscious precursor states, with the word order implicit in the precursor state. The brain models and remodels until the utterance is perfected in its analytic detail.

Order errors do occur in impaired verbal performance. Order errors in spelling (a specifically learned skill) characterize a particular syndrome of left-parietal impairment.[28] When people, normal or not, repeat arbitrary word sequences, they regularly make order errors. But word-order errors do not characterize neuropsychological impairments in naturally developed language.

In perception, again, one does not see "assembly errors." Percepts do not appear to be strung together, because brain-damaged patients do not report them to be misstrung (say, a cat with paws protruding from its head and its ears on its tail). Analogous regularities characterize sequential action. In ideational apraxia, in which the ability to execute familiar sequences is impaired, the action sequence becomes shortened by the omission of one or more components. But the components are not reversed or jumbled in their sequence.

Also eloquent in its absence from the literature is the partial disruption of a "mental model" of the surrounding world or of one's own body. When incoming information is obstructed by a lesion somewhere between the retina and primary visual cortex, patients have blind patches in their visual fields (scotomas). But they do not have holes in mental models; they do not complain of systematically being unable to discern objects in a particular egocentric relative location, however they look at their world. Deficit in a visual mental model should be apparent regardless of retinal angle of view, but such mental models do not appear to exist. The same reasoning applies to the so-called body schema. If there were a representation (a model) of the body parts in the brain, then focal damage should in one case implicate one part (the right elbow) and in another case, another part (the left hip). Putting together localizations from many patients, we could then assemble a cerebral topography for the body schema, like assembling a jigsaw puzzle. But such regional imperfections in the body image (partial asomatognosias) have not been reported.[29] As Andy Clark in this issue infers on other grounds, we do not build up models of the world and our bodies for reference purposes. Instead, since we carry our body and our world around with us, we simply refer to them when necessary.

The nonexistence of a deficit illuminates another issue of current interest: the alleged "binding" of visual features into perceived objects, of objects into scenes, scenes into episodes.[30] I assume that the brain can be stopped from doing anything it customarily does by an appropriately placed lesion. Some patients certainly report that objects look distorted (metamorphopsia) or persist in view (like afterimages) after the patient looks away (palinopia). But how would a patient describe his failure to bind, for instance, a color to a shape? Given the limitations of iconic memory, anyone might misremember which color belonged to which shape when briefly shown multiple, diversely colored, different shapes.[31] But primary failure to bind should express itself differently, as in "I see some colors and I see some shapes, but they do not go together." No such free floating of the colors of shapes or the shapes of colors is on record.

Another form of presumptive binding that lacks neuropsychological reality is cross-modal binding. No case had been reported in which within-modality perception pursued a normal course, but the percepts could not be combined across modalities. In a centered brain we would expect that there is convergence to a highest level in each modality, the activity of which incorporates the unified experience within that modality. But no such "summary module" is to be found.[32] On the contrary, it is the earliest-evolved cortical sensory areas that tend to be multimodal, and no single one of them encompasses all the modalities.[33] Earlier in brain evolution, the modality of input did not differentiate out. The trick is not to combine modalities, but to differentiate them.

The nonexistent deficits all send the same message: experience is not a composite assembled out of its parts. The contrary position—that experience is carved out of a less differentiated whole—gains plausibility. While no truly apt metaphor for how the brain works comes to mind, "crystallizing out" seems more fitting than "assembling together." Interestingly, brain development proceeds according to similar principles. The newborn has a full complement of neurons; further development proceeds by selective cell death and elimination of synaptic connections. The biological chisel prefigures the microgenesis of brain states.[34]

ATTENTION

The traditional model of brain organization in effect consigns attention to a central overseer—a central processing unit, central executive, a homunculus or femincula. This is unavoidable if information is integrated by multiple stepwise convergences to a central decision point (derisively pinpointed by William James as the "pontifical cell"). Decisions would then be implemented along a conversely diverging series of stages.[35] Carl Wernicke set the agenda a century ago: "To find the route, the telegraph line, by which the telegram is conveyed."[36] Attention would resemble Jeremy Bentham's synopticon—a neural "beam" that emanates from a central "lighthouse" and illuminates the action anywhere in the brain at will.

Contemporary neuropsychology steers us in a different direction. There is no forebrain lesion that selectively abolishes the ability to attend, across domains (nor one that selectively ablates consciousness).[37] Some lesions of the network result in altered behavior that the residual intact network takes cognizance of. Other lesions simply remove the affected domain from the sphere of consciousness, and the patient experiences neither function nor malfunction in that domain. It follows that the individual is aware of some types of deficits; of others he is not aware. He is aware, and complains vehemently, of difficulty with word finding (anomia), calculating (acalculia), recognizing faces (agnosia for faces), or printed words (alexia). The brain self-monitors, and mismatch between anticipated and actual outcome obtrudes into awareness.[38] In contrast, awareness is absent or incomplete when a right-posterior lesion causes "neglect" of the left side of space and person, leading to unawareness of blindness, paralysis, or anesthesia on the left.[39]

Deficits without awareness are distortions of attention. It is not that there is a gap or lacuna in attention; rather, attention is occupied elsewhere. In left neglect, attention and intention are biased rightward. When one attends selectively, one is largely oblivious of what remains unattended. This is even more pronounced with people who have attentional neuropathology. One can only be aware of how constricted one's attention is if one can internally represent (imagine) the possibility of attending some other way. This is done by initiating, but then aborting, such an attentional shift. If the brain substrate for such imaging is nonfunctional, one cannot experience the fact that one's attention is curtailed.

Remarkably, neglect patients' attentional capability can temporarily be normalized by the simple maneuver of stimulating the vestibular (balance) system by irrigating the opposite ear canal with lukewarm water.[40] Stimulating the vestibular system with water below body temperature directs ascending activation to the injured opposite hemisphere. It becomes clear that there had been an activation imbalance between hemispheres such that the uninjured hemisphere preempted attention, targeting it according to its natural bent to the opposite side of space and of the body. When the imbalance is corrected, neglect is no longer apparent. In this normalized state the neglect patient is not aware that his attention had previously been restricted. When the normalizing effect wears off, the patient again does not remember that his attention was different moments earlier.[41]

Unilateral neglect teaches us that awareness is a property of the ongoing pattern of forebrain activation, and its contents are those that the activated cell assemblies represent. I have called this controlling pattern of activation of the global network the "dominant focus."[42] Attending to a side is controlled by the reciprocal (negative feedback) interaction of a right-sided facility that directs selective orienting toward the left extreme of the viewed display, and a corresponding left-sided facility that directs it rightward.[43] If the right-sided facility is inactivated, attention that is now exclusively controlled by the disinhibited left hemisphere swings rightward—in any modality, in action and anticipation as well as in perception, and also in mental imagery. The patient is not aware that anything has gone wrong. He can no more think about attending to the left than do it (or imagine doing it).

I have referred to the distributed functioning outlined above as the "uncentered brain."[44] In the uncentered brain there is no location at which a lesion can abolish consciousness and yet leave intact the processes that compute what we are conscious of. When consciousness lapses, cognitive processes lapse. No lesion strips awareness from the functioning brain like a layer off an onion. Consciousness must be inherent in the working machinery; it cannot be its product.

Disorders of awareness restrict the contents of consciousness. The lesions that cause them are extensive; they distort and curtail the flexibility of the global network. But even limited lesions influence the functioning of the network as a whole.

LESION EFFECTS AT A DISTANCE

"Pure" or "specific" deficits are something of an abstraction. Lesions exert remote effects on destinations to which their sites are heavily connected. The downstream area can be deprived of excitation or released from inhibition. Moreover, any lesion can interfere with cognition in general. Patients perform less well than matched control subjects even on tasks that are quite unrelated to their specific deficits (as complex as problem solving or as simple as reacting to a flash or a click). Selective impairments are disproportionate rather than pure, and any impairment of the network handicaps the network as a whole. Modularity falls short.

When one hemisphere suffers major injury by stroke, the other hemisphere has been found to have become hypoactivated.[45] While ICA is in effect, not only the anesthetized hemisphere but to a lesser extent the opposite hemisphere also is underactivated, as judged by EEG slow waves and reduced blood flow tracked by a radioactive tracer (both of which indicate decreases in metabolic activity).[46] Even the opposite cerebellum is hypoactive during ICA. Each hemisphere must normally be activating the other side of the brain at a distance.

Striking distance effects are revealed by the dramatic and well publicized consequences of surgical section of the corpus callosum, the split-brain state, in which the two hemispheres are partly uncoupled. When stimulation is limited to one of the disconnected hemispheres and withheld from the other, the unstimulated hemisphere is typically unable to respond meaningfully (though there are exceptions). It is customary to invoke interrupted interhemispheric flow of information to explain hemispheric independence.[47] But much, if not all, split-brain phenomenology can alternatively be attributed to the loss of transcallosal cross-activation.[48] In the intact brain, the activated working hemisphere coactivates its partner, safeguarding its readiness to respond. Callosal section disrupts cross-activation, letting the unstimulated hemisphere lapse into sluggish unresponsiveness. In the intact brain, when all or much of one hemisphere is underactivated, this leads to some underactivation also of the other. This is not because both necessarily collaborate in tasks, but because when one is active the other must be ready also to be called upon at any time. The underactivation at a distance may not manifest itself behaviorally as a cognitive deficit but perhaps rather as a sluggishness on the part of the hemisphere to assume control when the need to do so comes its way.

Distance effects help the differentiated network maintain its precarious functional unity. Unity is not inherent in the network. It is secured by the neural architecture and dissolves in certain pathologies and under certain conditions, even in normals.[49]

AFFECT CONTROLS REASON

As was exemplified by unilateral neglect, when a specialized network is lesioned it may relinquish its role to another (opponent) network that propels behavior in the opposite direction. The symptoms that result are of two kinds. The first is those that derive from the loss of the role of the damaged or underactivated system; these are deficiencies, referred to as negative symptoms. The second is those that derive from the overactivity of the opponent system, which has been released from inhibition by the damaged network. It contributes positive symptoms. For instance, in neglect, unawareness of the left is a negative manifestation; excessive orienting to the right is a positive symptom. Many such opponent couplings are represented in the central nervous system. They underlie contrary behaviors that cannot be combined, between which the individual must choose. Sometimes the choice is automatic: turn left, to the door, and not right, into the wall. But often it requires cost accounting to determine the prospective benefits and costs of a given behavioral choice. This is where affect comes in.

Motivation drives behavior; absent motivation, the organism, however well equipped, remains inert. Positive emotions, such as elation, satisfaction, and gratification, earmark a particular action sequence that had a successful outcome as worth repeating in the future. Conversely, negative emotions attend mismatch between a goal and the actual outcome; this failure, highlighted subjectively by fear, revulsion, disgust, and other disagreeable feelings, calls for a halt in the action plan and a reappraisal of the situation. Attention broadens, additional factors are taken into consideration, and ideally a new improved strategy is devised. The action plan, narrowing towards the goal, is largely driven by the left hemisphere. The "suspend operation" instruction, accompanied by the surge of negatively charged arousal, is more right hemispheric in origin.[50]

Faulty monitoring of the outcomes of action plans may result in psychopathology. There might be an imbalance between the network that registers "match" and the network that registers "mismatch." A tendency to register spurious "match" outcomes would appear as a denial of deficit. Spurious "mismatch" results would send the individual back over and over again to repeat obsessively activities that others would consider completed. Failure to plan ahead (in left prefrontal dysfunction) would lead to an oppressive frequency of failure and interrupt experiences, engendering dysthymic moods and a sense of helplessness. A disinhibited arousal response to mismatch would present as mood swings or "hysterical overreaction." Conversely, too rigorous a correction of such arousal (inhibition of right posterior by right prefrontal cortex) would produce the emotional impoverishment and lack of capacity to enjoy (anhedonia) of major depression (melancholia). Cyclical swings of activational predominance of the left and right hemisphere (i.e., of the continue and interrupt function, respectively) could be related to the alternation of mania and depression in bipolar disease. In each case, a strategically located cortical injury could mimic a psychiatric disorder. It does so not by creating qualitatively abnormal behaviors but by causing one type of behavior to be used to excess, and its opposite not enough. In turn, the psychiatric disorder (whatever its cause) plays out in part by creating imbalance between cortical processing units, as described above. This imbalance is thought to result from imbalances between neurotransmitter systems.

The fact that a neural system is in place does not guarantee that it will participate in the control of behavior; it needs to be switched on, or activated. Powerful ascending activating systems hold the cortical units ready to respond. Different activating systems are powered by different neurotransmitters.[51] Either a deficiency in the action of a few neurotransmitters, or even a single one, or an excess in the action of another is held responsible or suspected for several psychopathologies. These chemical dysfunctions differ from structural lesions in that they can potentially be corrected with psychoactive drugs that either assist or impede the efficacy of particular neurotransmitters.

The overall lesson is that reason is the instrument of affect, not its master. Well-tuned affect enables reason to be deployed to best advantage. Both involve the highest cortical level of functioning. Poorly tuned affect compromises reason, and malfunctioning reason is reflected in deviant affect.

SCULPTING COGNITIONS

The notion that the brain is a switchboard for signals (or symbols), parts of which can be disassembled without consequence for the remaining net, is outdated, though persistent. The global net is depleted of resources by injury to one of its parts. It reorganizes to the extent possible to replace, or compensate for, the impaired capability. But now it operates with a restricted computational space, and this becomes apparent when the patient is asked to do something novel or difficult. In short, the injured brain operates on a skimpier neuronal and cognitive base, with fewer differentiated activity patterns available. In the absence of the specialized generator, another area may take over its role. The right hemisphere assumes language function when it is relinquished by the damaged left. The brain reacts as a whole to local impairment, demonstrating the connectedness of the differentiated neural net. The nonlinear dynamical model of the brain (discussed by Andy Clark in this issue) conceived in terms of coexisting attractor states—or better, in view of the transient nature and restless onrush of mental states, a trajectory through state space—seems to be consistent with evidence from lesion studies. The pattern of activation is global; its details are local.

During every conscious moment, the whole forebrain is a landscape (brainscape?) of peaks and valleys of activation (an energy topology). This pattern is as much the creation of the cell assemblies that are inhibited, and thus do not fire, as of those that contribute activation maxima. It updates Sherrington's fancy of the brain as a magical loom across which lights incessantly twinkle on and off.

The poet Stanley Kunitz remarked accurately, "We are not souls, but systems, and we move in clouds of our unknowing." We are beginning to know a little about ourselves as systems, and we are as interesting as Dyson led us to expect.

ENDNOTES

[1] Douglas Derryberry and Don M. Tucker, "The Adaptive Base of the Neural Hierarchy: Elementary Motivational Controls on Network Function," in P. Rakic and W. Singer, eds., *Neurobiology of Neocortex* (New York: Wiley, 1990).

[2] Marcel Kinsbourne and Elizabeth K. Warrington, "A Disorder of Simultaneous Form Perception," *Brain* 85 (1962): 461–468.

[3] Janet Jankoviak et al., "Preserved Visual Imagery and Categorization in a Case of Associative Visual Agnosia," *Journal of Cognitive Neuroscience* 4: 119–131. Martha J. Farah, *Visual Agnosia* (Cambridge, Mass.: MIT Press, 1990).

[4] Marcel Kinsbourne and Elizabeth K. Warrington, "A Case Showing Selectively Impaired Oral Spelling," *Journal of Neurology, Neurosurgery and Psychiatry* 29 (1965): 219–223; Marcel Kinsbourne and David Rosenfield, "Agraphia Selective for Written Spelling: An Experimental Case Study," *Brain and Language* 1 (1974): 215–225.

[5] Marcel Kinsbourne, "Behavioral Analysis of the Repetition Deficit in Conduction Aphasia," *Neurology* 22 (1972): 1126–1132; Morris Freedman, Michael P. Alexander, and Margaret A. Naeser, "Anatomic Basis of Transcortical Motor Aphasia," *Neurology* 34 (1984): 409–417.

[6] Marcel Kinsbourne and Bear Hiltbrunner, "A Selective Deficit in Cursive Writing" (unpublished).

[7] Claudio Luzzatti, R. Rumiati, and G. Ghirardi, "Visuo-verbal Disconnection and its Anatomical Constraints in Optic Aphasia," *Brain and Cognition* 32 (1996): 199–202.

[8] Marcel Kinsbourne and Elizabeth K. Warrington, "Observations on Colour Agnosia," *Journal of Neurology, Neurosurgery and Psychiatry* 27 (1964): 296–299.

[9]Marcel Kinsbourne, "Hemispheric Specialization and the Growth of Human Understanding," *American Psychologist* 37 (1982): 411–420.

[10]Lynn C. Robertson and Dean Delis, "Part-whole Processing in Unilateral Brain-damaged Patients: Dysfunction of Hierarchical Organization," *Journal of Neurosciences* 8 (1986): 3735–3769.

[11]Elizabeth K. Warrington, Merle James, and Marcel Kinsbourne, "Drawing Disability in Relation to Laterality of Cerebral Lesion," *Brain* 89 (1966): 53–82.

[12]Vadim Deglin and Marcel Kinsbourne, "Divergent Thinking Styles of the Hemispheres: How Syllogisms are Solved during Transitory Hemisphere Suppression," *Brain and Cognition* 31 (1996): 285–307.

[13]Gail L. Risse, J. R. Gates, and M. C. Fangman, "A Reconsideration of Bilateral Language Representation based on the Intracarotid Amobarbital Procedure," *Brain and Cognition* 33 (1997): 118–132.

[14]Marcel Kinsbourne, "The Minor Cerebral Hemisphere as a Source of Aphasic Speech," *Archives of Neurology* 25 (1971): 302–306; Christopher Code, *Language, Aphasia and the Right Hemisphere* (Chichester, England: Wiley, 1987). Marcel Kinsbourne, "The Right Hemisphere and Recovery from Aphasia," in Brigitte Stemmer and Harry A. Whitaker, eds., *Handbook of Neurolinguistics* (New York: Academic Press, 1997).

[15]Heiko Braak, "On Magnopyramidal Temporal Fields in the Human Brain— Probable Morphological Counterparts of Wernicke's Sensory Speech Region," *Anatomy and Embryology* 152 (1978): 141–169.

[16]Max S. Cynader et al., "General Principles of Cortical Organization," in Pasko Rakic and Wolf Singer, eds., *Neurobiology of the Neocortex* (New York: Wiley, 1988); Deepak N. Pandya, Benjamin Seltzer, and Helen Barbas, "Input-Output Organization of the Primate Cerebral Cortex," in Horst D. Steklis and J. Erwin, eds., *Comparative Primate Biology* (New York: Liss, 1988).

[17]Antonio R. Damasio, *Descartes' Error* (New York: Putnam, 1994).

[18]Karl H. Pribram, *Languages of the Brain* (New York: Prentice-Hall, 1971); P. W. Burgess and Tim Shallice, "Response Suppression, Initiation and Strategy Use following Frontal Lobe Lesions," *Neuropsychologia* 34 (1996): 263–273.

[19]Damasio, *Descartes' Error.*

[20]Deepak Pandya and C. L. Barnes, "Architecture and Connections of the Frontal Lobe," in Ellen Perecman, ed., *The Frontal Lobes Revisited* (New York: IRBN, 1987); Timothy W. Robbins, "Dissociating Executive Functions of the Prefrontal Cortex," *Philosophical Transactions of the Royal Society of London* B351 (1996): 1463–1472; Richard E. Passingham, "Attention to Action," *Philosophical Transactions of the Royal Society of London* B351 (1996): 1473–1479.

[21]Marcel Kinsbourne and Elizabeth K. Warrington, "Jargon Aphasia," *Neuropsychologia* 1 (1963): 27–37.

[22]Marcel Kinsbourne and Robert E. Hicks, "Functional Cerebral Space: A Model for Overflow, Transfer and Interference Effects in Human Perfor-

mance: A Tutorial Review," in Jean Requin, ed., *Attention and Performance VII* (Hillsdale, N.J.: Erlbaum, 1978).

[23]Marcel Kinsbourne, "Attentional Dysfunctions and the Elderly: Theoretical Models and Research Perspectives," in Leonard W. Poon et al., eds., *New Directions in Memory and Aging* (Hillsdale, N.J.: Erlbaum, 1980).

[24]Gordon G. Globus and Joseph P. Arpaia, "Psychiatry and the New Dynamics," *Biological Psychiatry* 35 (1994): 352–264; James P. Sutton and James A. Anderson, "Computational and Neurobiological Features of a Network of Networks," *Computation and Neural Systems* (forthcoming); Cynader et al., "General Principles of Cortical Organization"; Pandya et al., "Input-Output Organization of the Primate Cerebral Cortex."

[25]Cynader et al., "General Principles of Cortical Organization"; Pandya et al., "Input-Output Organization of the Primate Cerebral Cortex."

[26]Jason W. Brown, *The Life of the Mind: Selected Papers* (Hillsdale, N.J.: Erlbaum, 1988); Brian Kolb, "Brain Development, Plasticity and Behavior," *American Psychologist* 9 (1989): 1203–1212.

[27]John C. Marshal and Freda Newcombe, "Patterns of Paralexia: A Psycholinguistic Approach," *Journal of Psycholinguistic Research* 2 (1973): 175–199.

[28]Marcel Kinsbourne and Elizabeth K. Warrington, "Disorders of Spelling," *Journal of Neurology, Neurosurgery and Psychiatry* 27 (1964): 224–228.

[29]Marcel Kinsbourne, "Awareness of One's Own Body: A Neuropsychological Hypothesis," in Jose Luis Bermudez, Anthony J. Marcel, and Naomi Eilan, eds., *The Body and the Self* (Cambridge, Mass.: MIT Press, 1995).

[30]Francis Crick, "Function of the Thalamic Reticular Complex: The Searchlight Hypothesis," *Proceedings of the National Academy of Sciences* 81 (1984): 4586–4590; Wolf Singer, "Synchronization of Cortical Activity and its Putative Role in Information Processing and Learning," *Annual Review of Physiology* 55 (1993): 349–374.

[31]Anne M. Treisman and G. Gelade, "A Feature Integration Theory of Attention," *Cognitive Psychology* 12 (1980): 97–136.

[32]Semir Zeki and S. Shipp, "The Functional Logic of Cortical Connections," *Nature* 335 (1988): 311–317.

[33]Cynader et al., "General Principles of Cortical Organization"; Pandya et al., "Input-Output Organization of the Primate Cerebral Cortex."

[34]Brown, *The Life of the Mind*; Kolb, "Brain Development, Plasticity and Behavior."

[35]Norman Geschwind, "Disconnexion Syndromes in Animals and Man," *Brain* 88 (1965): 237–294, 585–644.

[36]Carl Wernicke, *Der Aphasische Symptomenkomplex* (Breslau, Germany: Cohn and Weigert, 1874).

[37]Marcel Kinsbourne, "Integrated Cortical Field Model of Consciousness," in Anthony J. Marcel and Eduardo Bisiach, eds., *The Concept of Consciousness in Contemporary Science* (London: Oxford University Press, 1988).

38Eran Zaidel, "Hemispheric Monitoring," in D. Otteson, ed., *Duality and Unity of the Brain* (London: McMillan, 1987).

39Marcel Kinsbourne, "Hemineglect and Hemisphere Rivalry," in Edwin A. Weinstein and Robert P. Friedland, eds., *Hemi-inattention and Hemisphere Specialization: Advances in Neurology* (New York: Raven, 1977).

40J. Silberpfennig, "Contributions to the Problem of Eye Movements: III, Disturbance of Ocular Movements with Pseuohemianopsia in Frontal Tumors," *Confinia Neurologica* 4 (1949): 1–13; Alan B. Rubens, "Caloric Stimulation and Unilateral Neglect," *Neurology* 35 (1985): 1019–1024.

41Vilayanur S. Ramachandran, "Anosognosia in Parietal Lobe Syndrome," *Consciousness and Cognition* 4 (1995): 22–51.

42Kinsbourne, "Integrated Cortical Field Model of Consciousness."

43Marcel Kinsbourne, "Lateral Interactions in the Brain" and "Mechanism of Hemispheric Interaction in Man," in Marcel Kinsbourne and W. Lynn Smith, eds., *Hemispheric Disconnection and Cerebral Function* (Springfield, Ill.: Thomas, 1974).

44Marcel Kinsbourne, "Models of Consciousness: Serial or Parallel in the Brain?" in Michael S. Gazzaniga, ed., *The Cognitive Neurosciences* (Cambridge, Mass.: MIT Press, 1995).

45R. J. Andrews, "Transhemispheric Diaschisis: A Review and Comment," *Stroke* 22 (1991): 943–949.

46David W. Loring et al., *Amobarbital Effects and Lateralized Brain Function: The Wada Test* (New York: Springer, 1992); D. McMakin et al., "Assessment of the Functional Effect of the Intracarotid Sodium Amytal Procedure Using Co-registered MRI/HMPAO-SPECT and SEEG," *Brain and Cognition* 33 (1997): 50–70.

47Eran Zaidel, Jeffrey M. Clarke, and B. Suyenobu, "Hemispheric Independence: A Paradigm Case for Cognitive Neuroscience," in Arnold B. Scheibel and Adam F. Wechsler, eds., *Neurobiology of Higher Cognitive Function* (New York: Guilford Press, 1990).

48Kinsbourne, "Lateral Interactions in the Brain"; Kinsbourne, "Mechanism of Hemispheric Interaction in Man"; Marcel Kinsbourne, "The Corpus Callosum as a Component of a Circuit for Selection," in Eran Zaidel, Marco Iacoboni, and A. Pascual-Leone, eds., *The Corpus Callosum in Sensory Motor Integration: Individual Differences and Clinical Applications* (New York: Plenum, 1998); Yves Guiard, "Cerebral Hemispheres and Selective Attention," *Acta Psychologica* 46 (1978): 41–61.

49Anthony Marcel, "Slippage in the Unity of Consciousness," in *Experimental and Theoretical Studies of Consciousness* (Ciba Foundation Symposium 174) (New York: Wiley, 1993), 168–186.

50Marcel Kinsbourne, "A Model of Adaptive Behavior Related to Cerebral Participation in Emotional Control," in Guido Gainotti and C. Caltagirone, eds., *Emotions and the Dual Brain* (New York: Springer, 1989).

51Timothy W. Robbins and Barry J. Everitt, "Arousal Systems and Attention," in Gazzaniga, ed., *The Cognitive Neurosciences*.

Andy Clark

Where Brain, Body, and World Collide

T HE BRAIN FASCINATES BECAUSE IT IS the biological organ of mindfulness itself. It is the inner engine that drives intelligent behavior. Such a depiction provides a worthy antidote to the once-popular vision of the mind as somehow lying outside the natural order. But it is a vision with a price, for it has concentrated much theoretical attention on an uncomfortably restricted space—the space of the inner neural machine, divorced from the wider world that then enters the story only via the hygienic gateways of perception and action. Recent work in neuroscience, robotics, and psychology casts doubt on the effectiveness of such a shrunken perspective. Instead, it stresses the unexpected intimacy of brain, body, and world and invites us to attend to the structure and dynamics of extended adaptive systems—ones involving a much wider variety of factors and forces. While it needs to be handled with some caution, I believe there is much to be learned from this broader vision. The mind itself, if such a vision is correct, is best understood as the activity of an essentially *situated* brain: a brain at home in its proper bodily, cultural, and environmental niche.

SOFTWARE

Humans, dogs, ferrets—these are, we would like to say, mindful things. Rocks, rivers, and volcanoes are not. And no doubt there are plenty of cases in between, such as insects or bacteria. In the

*Andy Clark is Professor of Philosophy and Director of the Philosophy/Neuroscience/
Psychology Program at Washington University in St. Louis.*

natural order, clear cases of mindfulness always involve creatures with brains. Hence, in part, the fascination of the brain: understanding the brain seems crucial to the project of understanding the mind. But how should such an understanding proceed?

An early sentiment—circa 1970, and no longer much in vogue—was that understanding the mind depended rather little on understanding the brain. The brain, it was agreed, was in some sense the physical medium of cognition. But everything that mattered about the brain qua mind-producing engine turned not on the physical details but on the computational and information-processing strategies that the neural stuff (merely) implemented. There was something importantly right about this view, but something desperately wrong as well.

What was right was the observation that understanding the physical workings of the brain was not going to be enough; we would need also to understand how the system was organized at a higher level in order to grasp the roots of mindfulness in the firing of neurons. This point is forcefully made by the cognitive scientist Brian Cantwell Smith, who draws a parallel with the project of understanding ordinary computer systems. Consider, for example, a standard personal computer running a tax-calculation program. We could quite easily answer all the "physiological" questions (using source code and wiring diagrams) while still lacking any real understanding of what the program does or even how it works.[1] To really understand how mental activity yields mental states, many theorists believe, we must likewise understand something of the computational/information-processing organization of the brain. Physiological studies may contribute to this understanding. But even a full physiological story would not, in and of itself, reveal how brains work qua mind-producing engines.[2]

The danger, of course, was that this observation could be used as an excuse to downplay or marginalize the importance of looking at the biological brain *at all*. And so it was that, in the early days of cognitive science, it was common to hear real neuroscience dismissed as having precious little to offer to the general project of understanding human intelligence. Such a dismissal, however, could not long be sustained. For although it is (probably) true to say that a computational understanding is in principle independent of the details of any specific implementation in hardware (or wetware),[3] the project of *discovering* the relevant computational description (especially for biological systems) is surely not.

One key factor is evolution. Biological brains are the product of biological evolution and as such often fail to function in the ways we (as human designers) might expect.[4] This is because evolution is both constrained and liberated in ways we are not. It is constrained to build its solutions incrementally via a series of simpler but successful ancestral forms. The human lung, to give one example, is built via a process of tinkering with the swim bladder of the fish.[5] The engineer might design a better lung from scratch. The tinkerer, by contrast, must take an existing device and subtly adapt it to a new role. From the engineer's ahistorical perspective, the tinkerer's solution may look bizarre. Likewise, the processing strategies used by biological brains may surprise the computer scientist. Such strategies have themselves been evolved via a process of incremental, piecemeal tinkering with older solutions.

More positively, biological evolution is *liberated* by being able to discover efficient but "messy" or unobvious solutions, ones that may, for example, exploit environmental interactions and feedback loops so complex that they would quickly baffle a human engineer. Natural solutions (as we shall later see) can exploit just about any mixture of neural, bodily, and environmental resources along with their complex, looping, and often nonlinear interactions. Biological evolution is thus able to explore a very different solution space (wider in some dimensions, narrower in others) than that which beckons to conscious human reason.

There are, of course, ways around this apparent mismatch. Computationalists lately exploit so-called "genetic algorithms" that roughly mimic the natural process of evolutionary search and allow the discovery of efficient but loopy and interactive adaptive strategies.[6] Moreover, hard neuroscientific data and evidence is increasingly available as a means of helping expand our imaginative horizons so as to appreciate better the way real biological systems solve complex problems. The one-time obsession with the "software level" is thus relaxing, in favor of an approach that grounds computational modeling in a more serious appreciation of the biological facts. In the next section I highlight some challenging aspects of such recent research—aspects that will lead us directly to a confrontation with the situated brain.

WETWARE, AND SOME ROBOTS

Recent work in cognitive neuroscience underlines the distance separating biological and "engineered" solutions to problems, and it displays an increasing awareness of the important interpenetration—in biological systems—of perception, thought, and action. Some brief examples should help fix the flavor.

As a gentle entry point, consider some recent work on the neural control of monkey finger motions. Traditional wisdom depicted the monkey's fingers as individually controlled by neighboring groups of spatially clustered neurons. According to this story, the neurons (in motor area 1, or M1) were organized as a "somatotopic map" in which a dedicated neural subregion governed each individual digit, with the subregions arranged in lateromedial sequence just like the fingers on each hand. This is a tidy, easily conceptualized solution to the problem of finger control. But it is the engineer's solution, not (it now seems) that of nature.

Marc Schieber and Lyndon Hibbard have shown that individual digit movements are accompanied by activity spread pretty well throughout the M1 hand area, and that precise, single-digit movements actually require *more* activity than some multidigit whole hand actions (such as grasping an object).[7] Such results are inconsistent with the hypothesis of digit-specific local neuronal groups. From a more evolutionary perspective, however, the rationale and design is less opaque. Some time earlier Schieber had conjectured that the basic case, from an evolutionary perspective, is the case of whole-hand grasping motions (used to grab branches, swing, acquire fruits, and so forth), and that the fundamental neural adaptations are thus geared to the use of simple commands that exploit inbuilt synergies of muscle and tendon so as to yield such coordinated motions.[8] The "complex" coordinated case is thus evolutionarily basic and neurally atomic. The "simple" task of controlling, for example, an individual digit represents the harder problem and requires more neural activity—viz., the use of some motor cortex neurons to *inhibit* the synergetic activity of the other digits. Precise single-digit movements require the agent to tinker with whole-hand commands, modifying the basic synergetic dynamics (of mechanically linked tendons, etc.) adapted to the more common task.

Consider next a case of perceptual adaptation. The human perceptual system can, we know—given time and training—adapt in quite powerful ways to distorted or position-shifted inputs. For example, subjects can learn how to coordinate vision and action while wearing lenses that invert the entire visual scene so that the world initially appears upside down. After wearing such lenses for a few days, the world is seen to flip over—various aspects of the world now appear to the subject to be in the normal upright position. Remove the lenses and the scene is again inverted until readaptation occurs.[9] William Thach and his colleagues at Washington University used a variant of such experiments to demonstrate the motor-specificity of some perceptual adaptations. Wearing lenses that shifted the scene *sideways* a little, subjects were asked to throw darts at a board. In this case, repeated practice led to successful adaptation, but of a motor loop–specific kind.[10] The compensation did not "carry over" to tasks involving the use of the nondominant hand to throw or to an underarm variant of the throw. Instead, adaptation looked to be restricted to a quite-specific combination of gaze angle and throwing angle: the one used in overarm, dominant-hand throwing.

Something of the neural mechanisms of such adaptation is now understood. It is known, for example, that the adaptation never occurs in patients with generalized cerebellar cortical atrophy, and that inferior olive hypertrophy leads to impaired adaptation. On the basis of this and other evidence, Thach and his coworkers speculated that a learning system implicating the inferior olive and the cerebellum (linked via climbing fibers) is active both in prism adaptation and in the general learning of patterned responses to frequently encountered stimuli. The more general lesson, however, concerns the nature of the perception-action system itself. It increasingly appears that the simple image of a general-purpose perceptual system delivering input to a distinct and fully independent action system is biologically distortive. Instead, perceptual and action systems work together, in the context of specific tasks, so as to promote adaptive success. Perception and action, on this view, form a deeply interanimated unity.

Further evidence for such a view comes from a variety of sources. Consider, for example, the fact that the primate visual system relies on processing strategies that are not strictly hierarchic but instead depend on a variety of top-to-bottom and side-to-side channels of influence. These complex inner pathways allow a combination of multiple types of information (high-level intentions, low-level perception, and motor activity) to influence all stages of visual processing. The macaque monkey (to take one well-studied example) possesses about thirty-two visual brain areas and over three hundred connecting pathways. The connecting pathways go both upwards and downwards (for example, from V1 to V2 and back again) and side to side (between subareas in V1).[11] Individual cells at "higher" levels of processing, such as visual area 4 (V4), do, it is true, seem to specialize in the recognition of specific geometric forms. But they will each also respond, to some small degree, to many other stimuli. These small responses, spread unevenly across a whole population of cells, can carry significant information. The individual cells thus function not as narrowly tuned single-feature detectors but as widely tuned filters reacting to a whole range of stimulus dimensions.[12] Moreover, the responses of such cells now look to be modifiable both by attention and by details of local task-specific context.[13]

More generally, back-projecting (corticocortical) connections tend, in the monkey, to outnumber forward-projecting ones; in other words, there are more pathways leading from deep inside the brain outwards towards the sensory peripheries than vice versa.[14] Visual processing may thus involve a variety of criss-crossing influences that could only roughly, if at all, be described as a neat progression through a lower-to-higher hierarchy.

Such complex connectivity opens up a wealth of organizational possibilities in which multiple sources of information combine to support visually guided action. Examples of such combinations are provided by Patricia Churchland, V. S. Ramachandran, and Terrence Sejnowski, who offer a neurophysiologically grounded account of what they term "interactive vision."[15] The interactive vision paradigm is there contrasted with approaches that assume a simple division of labor in which perceptual processing yields a rich, detailed inner representation of the visual scene—one that is later given as input to the reasoning and planning centers, which in turn calculate a course of action and send commands to the motor effectors. This simple image (of what roboticists call a "sense-think-act" cycle) is, it now seems, not true to the natural facts. In particular, daily agent-environment interactions often do not require the construction and use of detailed inner models of the full visual scene. Moreover, low-level perception may "call" motor routines that yield *better perceptual input* and hence improve information pickup. Beyond this, real-world actions may sometimes play an important role in the computational process itself. Finally, the internal representation of worldly events and structures may be less like a passive data structure or description and more like a direct recipe for action.

Evidence for the first of these propositions comes from a series of experiments in which subjects watch pictorial images on a computer screen. Then, as they continue to saccade around the scene (focusing first on one area, then another), small changes are made to the currently unattended parts of the display. The changes are made during the visual saccades. It is an amazing fact that, for the most part, quite large changes go unnoticed: changes such as the replacement of a tree by a shrub, or the addition of a car, the deletion of a hat, and so on.[16] Why do such gross alterations remain undetected? A compelling hypothesis is that the visual system is not even attempting to build a rich, detailed model of the current scene but is instead geared to using frequent saccades to retrieve information *as and when it is needed* for some specific problem-solving purpose. This fits nicely with Yarbus's classic finding that the pattern of such saccades varies (even with identical scenes) according to the type of task the subject has been set (for example, to give the ages of the people in a picture, to guess the activity they have been engaged in, and so on).[17] One explanation of our subjective impression of forming a rich inner representation of the whole scene is that we are prone to this illusion because we are able to perform fast saccades, retrieving information as and when required. (An analogy: a modern store may present the illusion of having a massive amount of goods stocked on the premises, because it always has what you want when you want it. But modern computer ordering systems can automatically count off sales and requisition new items so that the necessary goods are available just when needed and barely a moment before. This fine-tuned ordering system offers a massive saving of on-site storage while tailoring supply directly to customer demand.)[18]

Contemporary research in robotics avails itself of these same economies. Rodney Brooks, one of the pioneers of "new robotics," coined the slogan "the world is its own best model" to capture just this flavor.[19] A robot known as Herbert, to take just one example, was designed to collect soft-drink cans left around a crowded laboratory. But instead of requiring powerful sensing capacities and detailed advance planning, Herbert got by very successfully using a collection of coarse sensors and simple, relatively independent, behavioral routines. Basic obstacle avoidance was controlled by a ring of ultrasonic sensors that brought the robot to a halt if an object was in front of it. General locomotion (randomly directed) was interrupted if Herbert's simple visual system detected a rough tablelike outline. At this point a new routine kicks in and the table surface is swept using a laser. If the outline of a can is detected, the whole robot rotates until the can is centered in its field of vision. This simple physical action simplifies the pickup procedure by creating a standard action-frame in which a robot arm, equipped with simple touch sensors, gently skims the table surface dead ahead. Once a can is encountered, it is grasped and collected, and the robot moves on. Notice, then, that Herbert succeeds without using any conventional planning techniques and without creating and updating any detailed inner model of the environment. Herbert's "world" is composed of undifferentiated obstacles and rough tablelike and canlike outlines. Within this world the robot also exploits its own bodily actions (rotating the "torso" to center the can in its field of view) so as to simplify greatly the computational problem involved in eventually reaching for the can. Herbert is thus a simple example of the second proposition above—a system that succeeds using minimal representational resources, and one in which gross motor activity helps streamline a perceptual routine.[20]

The interactive vision framework envisages a more elaborate natural version of this same broad strategy, viz., the use of a kind of perceptuomotor loop whose role is to make the most of incoming perceptual information by combining multiple sources of information. The idea here is that perception is not a passive phenomenon through which motor activity is initiated only at the endpoint of a complex process whereby the animal creates a detailed representation of the perceived scene. Instead, perception and action engage in a kind of incremental game of tag, in which motor assembly begins long before sensory signals reach the top level. Thus, early perceptual processing may yield a kind of protoanalysis of the scene, enabling the creature to select actions (such as head and eye movements) whose role is to provide a slightly upgraded sensory signal. That signal may, in turn, yield a new protoanalysis indicating further visuomotor action, and so on. Even whole-body motions may be deployed as part of this process of improving perceptual pickup. Foveating an object can, for example, involve motion of the eyes, head, neck, and torso. Churchland and her associates put it well: "Watching Michael Jordan play basketball or a group of ravens steal a caribou corpse from a wolf tends to underscore the integrated, whole-body character of visuomotor coordination."[21] This integrated character is consistent with neurophysiological and neuroanatomical data that show the influence of motor signals in visual processing. There are—to take just two small examples—neurons sensitive to eye position in V1, V3, and LGN (lateral geniculate nucleus), and cells in V1 and V2 that seem to know in advance about planned visual saccades (i.e., they show enhanced sensitivity to the target).[22]

Moving on to the third proposition—that real-world actions may sometimes play an important role in the computational process itself—consider the task of distinguishing a figure from the ground (the rabbit from the field, or whatever). It turns out that this problem is greatly simplified using information obtained from head movement during eye fixation. Likewise, depth perception is greatly simplified using cues obtained by the observer's own self-directed motion. As the observer moves, close objects will show more relative displacement than farther ones. That is probably why, as Churchland, Ramachandran, and Sejnowski observe, head-bobbing behavior is frequently seen in animals: "A visual system that integrates across several glimpses to estimate depth has computational savings over one that tries to calculate depth from a single snapshot."[23]

And so, with the last proposition—that the neural representation of worldly events may be less like a passive data structure and more like a recipe for action—the driving force, once again, is computational economy. If the goal of perception and reason is to guide action (and it surely is, evolutionarily speaking), it will often be simpler to represent the world in ways rather closely geared to the kinds of actions we want to perform. To take a simple example, an animal that uses its visual inputs to guide a specific kind of reaching behavior (so as to acquire and ingest food) need not form an object-centered representation of the surrounding space. Instead, a systematic metrical transformation (achieved by a point-to-point mapping between two topographic maps) may transform the visual inputs directly into a recipe for reaching out and grabbing the food. In such a setup, the animal does not need to do any computational work on an action-neutral inner model so as to plan a reaching trajectory. The perceptual processing is instead tweaked, at an early stage, in a way dictated by the particular use to which the visual input is dedicated. This strategy is described in detail in Paul Churchland's account of the "connectionist crab," in which research in artificial neural networks is applied to the problem of creating efficient point-to-point linkages between deformed topographic maps.[24]

In a related vein, Maja Mataric of the Massachusetts Institute of Technology's Artificial Intelligence Laboratory has developed a neurobiologically inspired model of how rats navigate their environments. This model exploits the kind of layered architecture also used in the robot Herbert.[25] Of most immediate interest, however, is the way Mataric's robot learns about its surroundings. As it moves around a simple maze, it detects landmarks that are registered as a combination of sensory input and current motion. A narrow corridor thus registers as a combination of forward motion and short lateral distance readings from sonar sensors. Later, if the robot is required to find its way back to a remembered location, it retrieves an interlinked body of such combined sensory and motor readings.[26] The stored "map"of the environment is thus immediately fit to act as a recipe for action, since the motor signals are part of the stored knowledge. The relation between two locations is directly encoded as the set of motor signals that moved the robot from one to the other. The inner map is thus *itself* the recipe for the necessary motor actions. By contrast, a more classical approach would first generate a more objective map, which would then need to be *reasoned over* in order to plan the route.

The Mataric robot and the connectionist crab exemplify the attractions of what I call "action-oriented representations," representations that describe the world by depicting it in terms of possible actions.[27] This image fits in nicely with several of the results reported earlier, including the work on monkey finger control and the motor-loop specificity of "perceptual" adaptation. The products of perceptual activity, it seems, are not always action-neutral descriptions of external reality. They may instead constitute direct recipes for acting and intervening. We thus glimpse something of the shape of what some have described as a framework that is "motocentric" rather than "visuocentric."[28]

As a last nod in this same direction, consider the fascinating case of so-called mirror neurons.[29] These are neurons found in the monkey ventral premotor cortex that are action-oriented, context-dependent, and implicated in both self-initiated activity and passive perception. They are active both when the monkey observes a specific action (such as someone grasping a food item) and when the monkey performs the same action, where sameness implies not mere grasping but the grasping of a food item.[30] The implication, according to the psychologist and neuroscientist Marc Jeannerod, is that "the action. . .to be initiated is stored in terms of an action code, not a perceptual one."[31]

Putting all this together suggests a much more integrated model of perception, cognition, and action. Perception is itself tangled up with possibilities for action and is continuously influenced by cognitive, contextual, and motor factors. It need not yield a rich, detailed, and action-neutral inner model awaiting the services of "central cognition" so as to deduce appropriate actions. In fact, these old distinctions (between perception, cognition, and action) may sometimes obscure, rather than illuminate, the true flow of effect. In a certain sense, the brain is revealed not as (primarily) the engine of reason or quiet deliberation, but as the organ of *environmentally situated control*. Action, not truth and deductive inference, is the key organizing concept. This perspective, however, begs to be taken a step further. For if brains are best understood as controllers of environmentally situated activity, then might it not be fruitful to locate the neural contribution as just one (important) element in a complex causal web, spanning brains, bodies, and world? This potential gestalt shift is the topic of the next section.

WIDEWARE

Let us coin a term, "wideware," to refer to states, structures, or processes that satisfy two conditions. First, the item in question must be in some intuitive sense environmental; it must not, at any rate, be realized within the biological brain or the central nervous system. Bodily aspects and motions, as well as truly external items such as notebooks and calculators, thus fit the bill. Second, the item (state, structure, process) must play a functional role as part of an *extended cognitive process*: a process geared to the promotion of adaptive success via the gathering and use of knowledge and information, and one that loops out in some nontrivial way, so as to include and exploit aspects of the local bodily and environmental setting.

Of course, even what is intuitively a fully internalized cognitive process will usually involve contact with the external world. That much is demanded by the very ideas of knowledge acquisition and information-gathering. The notion of wideware aims to pick out instead a somewhat more restricted range of cases—ones in which the *kind* of work that cognitive science has typically assigned to the inner workings of the brain is at least partly carried out by processes of storage, search, and transformation, realized using bodily actions and/or a variety of external media. Understanding the human mind, I shall argue, will require us to attend much more closely to the role of such bodily actions and external media than was once anticipated.

To better fix the notion of wideware itself, consider first a very simple case involving "bodily backdrop." Computer-controlled machines are sometimes used to fit small parts into one another. The error tolerance here is very low; sometimes the robot arm will fail to make a match. A computationally expensive solution exists in which the control system includes multiple feedback loops that signal such failures and prompt the machine to try again in a minutely different orientation. It turns out, however, that a much cheaper, more robust, and more efficient procedure is simply to mount the assembler arms on rubber joints that "give" along two spatial axes. This bodily backdrop allows the control device to dispense with the complex feedback loops. Thanks to the rubber mountings, the parts "jiggle and slide into place just as if millions of tiny feedback adjustments to a rigid system were being continuously computed."[32]

Mere bodily backdrop, however, does not really count as an instance of an extended cognitive process. The computational and information-based operations are reduced, courtesy of the brute physical properties of the body. But they are not themselves extended *into* the world. Genuine cognitive and computational extension requires, by contrast, that the external or bodily operations are themselves usefully seen as performing cognitive or information-processing operations. A simple phototropic (light-following) robot—to take another negative example—does not constitute an extended cognitive or computational system. Although the presence of some external structure (light sources) is here vital to the robot's behavioral routines, those external structures are not doing the kind of work that cognitive science and psychology has typically assigned to inner neural activity. They are not acting so as to manipulate, store, or modify the knowledge and information that the organism uses to reach its goals. Sometimes, however, external structures and bodily operations do seem to be proper parts of the cognitive and computational processes themselves. Thus consider the use, in recent interactive vision research, of so-called deictic pointers. A classical (nondeictic) pointer is an inner state that can figure in computational routines and that can also "point" to further data structures. Such pointing allows for both the retrieval of additional information and the binding of the content of one memory location to another. One use of classical pointers is thus to bind information about spatial location ("top left corner of the visual field") to information about current features ("bright yellow mug"), yielding complex contents (in this case, "bright yellow mug in the top left corner of the visual field").[33] Since neural processing involves more or less distinct channels for object properties, object location, and object motion (for example), binding is an important element in neural computation. But binding, it now seems, can sometimes be achieved by the use of actual bodily orientations instead of linking inner-data structures.

The story is complex, but the basic idea is straightforward. It is to set up a system so that bodily orientations (such as saccadic eye motions leading to object fixation) directly yield the kinds of benefits associated with classical binding. Dana Ballard and colleagues have shown how to use visual scene fixation to directly associate the features represented (the properties of the perceived object) with an external spatial location.[34] Another example is the use of so-called "do-it-where-I'm-looking" processing routines in which a bodily motion (say, grasping) is automatically directed to whatever location happens to be currently fixated in the visual field. In all these cases, the authors comment: "The external world is analogous to computer memory. When fixating a location the neurons that are linked to the fovea refer to information computed from that location. Changing gaze is analogous to changing the memory reference in a silicon computer."[35]

Deictic binding provides a clear example of a case in which bodily motion does the *kind* of work that cognitive science has typically assigned only to inner neural activity. The deictic strategies use a combination of inner computation and gross bodily action to support a type of functionality once studied as an exclusive property of the inner neural system. Taking this perspective a step further, we next consider the use of external media as both additional memory and as potent symbol-manipulating arenas.

Portions of the external world, it is fairly obvious, often function as a kind of extraneural memory store. We may deliberately leave a film on our desk to remind us to take it for developing. Or we may write a note ("develop film") on paper and leave that on our desk instead. As users of words and texts, we command an especially potent means of off-loading data and ideas from the biological brain onto a variety of external media. This trick, I think, is not to be underestimated, for it affects not just the quantity of data at our command but also the kinds of operation we can bring to bear on it. Words, texts, symbols, and diagrams often figure *deeply* in the problem-solving routines developed by biological brains nurtured in language-rich environmental settings. Human brains, trained in a sea of words and text, will tend to develop computational strategies that directly factor in the reliable presence of a variety of external props and aids. The inner operations will then complement, but not replicate, the special manipulative potentials provided by the external media.

Consider, for example, the process of writing an academic paper.[36] You work long and hard, and at the end of the day you are happy. Being a good physicalist, you assume that all the credit for the final intellectual product belongs to your brain, the seat of human reason. But you are too generous by far. What really happened was (perhaps) more like this: The brain supported some rereading of old texts, materials, and notes. While rereading these, it responded by generating a few fragmentary ideas and criticisms. These ideas and criticisms were then stored as more marks on paper, in margins, on computer disks, or some other recording means. The brain then played a role in reorganizing this data on clean sheets, adding new on-line reactions and ideas. The cycle of reading, responding, and external reorganization is repeated, again and again. Finally, there is a product—a story, argument, or theory. But this intellectual product owes a lot to those repeated loops into the environment. Credit belongs to the agent-in-the-world. The biological brain is just a part (albeit a crucial and special part) of a spatially and temporally extended process, involving lots of extraneural operations, whose joint action creates the intellectual product. There is thus a real sense (or so I would argue) in which the notion of the "problem-solving engine" is really the notion of the *whole caboodle*: the brain and body operating within an environmental setting.

Consider, by way of analogy, the idea of a swimming machine; in particular, consider the bluefish tuna.[37] The tuna is paradoxically talented. Physical examination suggests it should not be able to achieve the aquatic feats of which it is demonstrably capable. It is physically too weak (by about a factor of 7) to swim as fast as it does, to turn as compactly as it does, or to move off with the acceleration it does. The explanation (according to the fluid dynamicists Michael and George Triantafyllou) is that these fish actively create and exploit additional sources of propulsion and control in their watery environments. For example, the tuna use naturally occurring eddies and vortices to gain speed, and they flap their tails so as actively to create additional vortices and pressure gradients that they then exploit for quick takeoffs and similar feats. The real swimming machine, I suggest, is thus the fish *in its proper context*: the fish plus the surrounding structures and vortices than it actively creates and then maximally exploits. The *cognitive machine,* in the human case, looks similarly extended.[38] We actively create and exploit multiple linguistic media, yielding a variety of contentful structures and manipulative opportunities whose reliable presence is then factored deep into our problem-solving strategies.

IMPLICATIONS

Software, wetware, and wideware, if our story is to be believed, form a deeply interanimated triad. The computational activities of the brain will be heavily sculpted by its biological "implementation." And there will be dense complementarity and cooperation between neural, bodily, and environmental forces and factors. What the brain does will thus be precisely fitted to the range of complementary operations and opportunities provided by bodily structure, motion, and the local environment. In the special case of human agency, this includes the humanly generated "whirlpools and vortices" of external, symbol-laden media: the explosion of wideware made available by the ubiquitous devices of language, speech, and text. The picture that emerges is undeniably complex. But what does it really mean both for our understanding of ourselves and for the practice of scientific inquiry about the mind?

Certain implications for our vision of ourselves are clear. We must abandon the image of ourselves as essentially disembodied reasoning engines. And we must do so not simply by insisting that the mental is fully determined by the physical, but by accepting that we are beings whose neural profiles are profoundly geared so as to press maximal benefit from the opportunities afforded by bodily structure, action, and environmental surroundings. Biological brains are, at root, controllers of embodied action. Our cognitive profile is *essentially* the profile of an embodied and situated organism. Just how far we should then press this notion of cognitive extension, however, remains unclear. Should we just think of ourselves as cognitive agents who co-opt and exploit surrounding structures (say, pen and paper) so as to expand out problem-solving capacities? Or is there a real sense in which the cognitive agent (as opposed to the bare biological organism) is thus revealed as an extended entity incorporating brain, body, and some aspects of the local environment? Normal usage would seem to favor the former. But the more radical interpretation is not as implausible as it may initially appear.

Consider an example. Certain Alzheimer's sufferers maintain an unexpectedly high level of functioning within the normal community. These individuals should not—given their performance on a variety of standard tests—be capable of living as independently as they do. Their unusual success is explained only when they are observed in their normal home environments, in which an array of external props and aids turns out to serve important cognitive functions.[39] Such props and aids may include the extensive use of labels (on rooms, objects, and so forth), the use of a "memory book" containing annotated photos of friends and relatives, the use of a diary for tasks and events, and simple tactics such as leaving all important objects (say, one's checkbook) in open view so as to aid retrieval when needed.[40] The upshot, clearly, is an increased reliance on various forms of wideware (or "cognitive scaffolding") as a means of counterbalancing a neurally based deficit.[41] But the pathological nature of the case is, in a sense, incidental. Imagine a whole community whose linguistic and cultural practices evolved so as to counterbalance a normal cognitive profile (within that community) identical to that of these individuals. The external props could there play the same functional role (of complementing a certain neural profile) but without any overtone of pathology-driven compensation. Finally, reflect that our own community is just like that. Our typical neural profile is different, to be sure, but *relative to that profile* the battery of external props and aids—laptops, filofaxes, texts, compasses, maps, slide rules—plays just the same role. They offset cognitive limitations built into the basic biological system.

Now ask yourself what it would mean, in the case of the Alzheimer's sufferer, to maliciously damage that web of external cognitive support. Such a crime has, as Daniel Dennett once remarked, much of the flavor of a harm to the *person,* rather than simple harm to property. The same may well be true in the normal case: deliberate theft of the poet's laptop is a very special kind of crime. Certain aspects of the external world, in short, may be so integral to our cognitive routines as to count as *part of the cognitive machinery* itself (just as the whorls and vortices are, in a sense, part of the swimming machinery of the tuna). It is thus something of a question whether we should see the cognizer as the bare biological organism (that exploits all those external props and structures), or as the organism-plus-wideware. To adopt the latter perspective is to opt for a kind of "extended phenotype" view of the mind, in which the relation between the biological organism and the wideware is as important and intimate as that of the spider and the web.[42]

The implications for the specific study of the mind are, fortunately, rather less ambiguous. The vision of the brain as a controller of embodied and situated action suggests the need to develop new tools and techniques capable of investigating the brain (literally) in action—playing its part in problem-solving routines in (as far as possible) a normal ecological context. Very significant progress has already been made, of course. Noninvasive scanning techniques, such as positron emission tomography (PET) scanning and magnetic resonance imaging (MRI), represent a giant leap beyond the use of single-cell recordings from anaesthetized animals. But we should not underestimate the distance that remains to be covered. For example, experimenters recently carried out some neural recordings from a locust in free flight.[43] This kind of ecologically realistic study is clearly mandated by the kinds of consideration we have put forward. Yet such investigations remain problematic due to sheer technological limitations. In the locust case, the researchers relied upon tiny radio transmitters implanted in the insect. But the information pickup remained restricted to a scanty two channels at a time.

Moreover, it is not just the information-gathering techniques that need work; the analytic tools we bring to bear on the information gathered need improving as well. If we take seriously the notion that brain, body, and world are often united in an extended problem-solving web, we may need to develop analytic and explanatory strategies that better reflect and accommodate this dense interanimation. To this end, there are (so far) two main proposals on the table. One proposal is to use the tools of dynamical systems theory. This is a well-established mathematical framework for studying the temporal evolution of states within complex systems.[44] Such a framework appeals, in part, because it allows us to use a single mathematical formalism to describe both internal and external organizations and (hence) allows us to treat complex looping interactions (ones that crisscross brain, body, and world) in a deeply unified manner. The other proposal is to take the kinds of analysis we traditionally applied only to the inner states and extend it outwards.[45] This means sticking with talk of representations and computations but applying these ideas to extended organizations encompassing, for example, multiple individuals, maps, texts, and social institutions. (My own view, which I will not defend here, is that we need to combine the two approaches by defining new, dynamical, process-based ways of understanding key terms like "representation" and "computation."[46])

Finally, the pervasive notion of a neural code or codes is now in need of major overhaul. If the notion is to apply to real biological systems, it must be relieved of a good deal of excess baggage. Natural neural codes, as we saw above, will often be closely geared to the particular details of body and world. The coding for single-finger motion is unexpectedly baroque, courtesy of the need to combat basic synergies created by the system of mechanically linked tendons. Other codes (for example, for whole-hand grasping motions) prove unexpectedly simple and direct. Moreover, there is evidence of the widespread use of task-specific, motor-oriented, and context-dependent encodings. Mirror neurons code for context-specific actions (say, grasping food) and function both in passive perceptions and active grasping. And interactive vision routines press large benefits from minimal and often task-specific forms of internal encoding. In all these cases, we discern a much more austere and action-oriented vision of neural encoding—a vision stripped of the excess baggage of a single, rich, languagelike inner code.[47]

The simple, almost commonsensical notion of the brain as a system that evolved in order to guide the actions of embodied agents in rich real-world settings thus yields substantial and sometimes challenging fruit. Gone is the vision of the environment as simply a source of problem-specifying inputs and an arena for action. Instead, both basic and imposed aspects of environmental order and complexity now emerge as fundamental components of natural problem-solving behavior. Gone too is the vision of the human brain as an organ of pure reason. In its place, we encounter the brain as a locus of action-oriented and activity-exploiting problem-solving techniques, and as a potent generator and exploiter of cognition-enhancing wideware. Taking this vision seriously, and turning it into a concrete and multidisciplinary research program, presents an exciting new challenge for the sciences of the mind.[48]

ENDNOTES

[1]See Brian Cantwell Smith, *On the Origin of Objects* (Cambridge, Mass.: MIT Press, 1996), 148. Note that Smith's worry, at root, concerns the gap between physiological and semantic or intentional questions.

[2]Thus we read, for example, that computational approaches make possible "a science of structure and function divorced from material substance [that] . . . can answer questions traditionally posed by psychologists." Zenon W. Pylyshyn, ed., *The Robot's Dilemma: The Frame Problem in Artificial Intelligence* (Norwood, U.K.: Ablex, 1987), 68.

[3]For example, see David Marr's distinction between the levels of computation, algorithm, and implementation. David Marr, *Vision* (San Francisco, Calif.: W. H. Freeman, 1982). For a powerful critique of the tendency of cognitive science to discount the importance of neuroscience, see George N. Reeke, Jr., and Gerald M. Edelman, "Real Brains and Artificial Intelligence," *Dædalus* 117 (1) (Winter 1988): 143–173.

[4]See, for example, Herbert Alexander Simon, *Models of Bounded Rationality*, 2 vols. (Cambridge, Mass.: MIT Press, 1982); Richard Dawkins, *The Extended Phenotype: The Gene as the Unit of Selection* (Oxford: Oxford University Press, 1982); and Andy Clark, *Being There: Putting Brain, Body, and World Together Again* (Cambridge, Mass.: MIT Press, 1997), chap. 5.

[5]François Jacob, "Evolution and Tinkering," *Science* 196 (1977): 1161–1166.

[6]For a review, see Clark, *Being There*, chap. 5.

[7]Marc Schieber and Lyndon Hibbard, "How Somatotopic is the Motor Cortex Hand Area?" *Science* 261 (1993): 489–492.

[8]Marc Schieber, "How might the Motor Cortex Individuate Movements?" *Trends in Neuroscience* 13 (11) (1990): 440–444. The notion of synergy aims to capture the idea of links that constrain the collective unfolding of a system comprising many parts. For example, the front wheels of a car exhibit a built-in synergy that allows a single driver "command" (at the steering wheel) to affect them both at once. Synergetic links may also be learned, as when we acquire an automated skill, and may be neurally as well as brute-physiologically grounded. See J. A. Scott Kelso, *Dynamic Patterns* (Cambridge, Mass.: MIT Press, 1995), 38, 52.

[9]For a survey of such experiments, see Robert B. Welch, *Perceptual Modification: Adapting to Altered Sensory Environments* (New York: Academic Press, 1978).

[10]In this case, *without* any perceived shift in the visual scene. See W. Thach, H. Goodkin, and J. Keating, "The Cerebellum and the Adaptive Coordination of Movement," *Animal Review of Neuroscience* 15 (1992): 403–442.

[11]See, for example, Daniel Felleman and David C. Van Essen, "Distributed Hierarchical Processing in the Primate Visual Cortex," *Cerebral Cortex* 1 (1991): 1–47.

[12]See David C. Van Essen and Jack Gallant, "Neural Mechanisms of Form and Motion Processing in the Primate Visual System," *Neuron* 13 (1994): 1–10.

[13]James Knierim and David C. Van Essen, "Visual Cortex: Cartography, Connectivity, and Concurrent Processing," *Current Opinion in Neurobiology* 2 (1992): 150–155. In a somewhat related vein, Caminiti and his coworkers showed that the directional preference of individual cells encoding reaching movement commands varies according to the initial arm position. R. Caminiti et al., "Shift of Preferred Directions of Premotor Cortical Cells with Arm Movements Performed Across the Workspace," *Experimental Brain Research* 82 (1990): 228–232. See also the discussion in Marc Jeannerod, *The Cognitive Neuroscience of Action* (Oxford: Blackwell, 1997), chap. 2.

[14]Though much of the connectivity is reciprocal. See David Van Essen and Charles Anderson, "Information Processing Strategies and Pathways in the Primate Retina and Visual Cortex," in Stephen Zornetzer, Joel L. Davis, and Clifford Lau, eds., *An Introduction to Neural and Electronic Networks* (New York: Academic Press, 1990). See also Patricia Churchland, V. S. Ramachandran, and Terence Sejnowski, "A Critique of Pure Vision," in Christof Koch and Joel Davis, eds., *Large-Scale Neuronal Theories of the Brain* (Cambridge, Mass.: MIT Press, 1994), 40.

[15]Churchland, Ramachandran, and Sejnowski, "A Critique of Pure Vision."

[16]The exception is if subjects are told in advance to watch out for changes to a certain feature. G. W. McConkie, "Where Vision and Cognition Meet," paper presented at the H.F.S.P. Workshop on Object and Sense Perception, Leuven, Belgium, 1990. See also Churchland, Ramachandran, and Sejnowski, "A Critique of Pure Vision."

[17]Alfred L. Yarbus, *Eye Movements and Vision* (New York: Plenum Press, 1967).

[18]See Churchland, Ramachandran, and Sejnowski, "A Critique of Pure Vision," and Dana H. Ballard, "Animate Vision," *Artificial Intelligence* 48 (1991): 57–86. My thanks to David Clark for pointing out the store analogy.

[19]See, for example, Rodney Brooks, "Intelligence Without Representation," *Artificial Intelligence* 47 (1991): 139–159.

[20]Jonathan Connell, "A Colony Architecture for an Artificial Creature," *MIT AI Lab Tech Report* 2 (5) (1991): 1.

[21]Churchland, Ramachandran, and Sejnowski, "A Critique of Pure Vision," 44.

[22]R. Wurtz and C. Mohler, "Enhancement of the Visual Response in the Monkey Striate Cortex and Frontal Eye Fields," *Journal of Neurophysiology* 39 (1976): 766–772.

[23]Churchland, Ramachandran, and Sejnowski, "A Critique of Pure Vision," 51.

[24]The quote is from Paul M. Churchland, *The Neurocomputational Perspective* (Cambridge, Mass.: MIT Press/Bradford Books, 1989). For an accessible introduction to artificial neural networks, see Paul Churchland, *The Engine of Reason, the Seat of the Soul* (Cambridge, Mass.: MIT Press, 1995).

[25]This is known as a "subsumption" architecture, because the layers each constitute a complete behavior-producing system and interact only in simple ways such as by one layer subsuming (turning off) the activity of another or by one layer co-opting and hence "building in" the activity of another. See Brooks, "Intelligence Without Representation."

[26]The retrieving is accomplished by a process of spreading activation among landmark encoding nodes. See Maja Mataric, "Navigating with a Rat Brain: A Neurobiologically Inspired Model for Robot Spatial Representation," in Jean-Arcady Meyer and Stewart W. Wilson, eds., *From Animals to Animats: Proceedings of the First International Conference on Simulation of Adaptive Behavior* (Cambridge, Mass.: MIT Press, 1991).

[27]Clark, *Being There*, 47. The Mataric robot is based on actual rat neurobiology; see B. McNaughton and L. Nadel, "Hebb-Marr Networks and the Neurobiological Representation of Action in Space," in *Neuroscience and Connectionist Theory*, ed. Mark A. Gluck and David E. Rumelhart (Hillsdale, N.J.: Erlbaum Associates, 1990). Action-oriented representations bear some resemblance to what the ecological psychologist J. J. Gibson called "affordances." Yet Gibson himself would reject our emphasis on inner states and encodings, for an affordance (as he sees it) is the potential of use and activity that the local environment offers to a specific kind of being: chairs afford sitting (to humans), and so on. J. J. Gibson, *The Ecological Approach to Visual Perception* (Boston, Mass.: Houghton Mifflin, 1979). The philosopher Ruth Millikan has developed a nice account of action-oriented representation under the label "pushmi-pullyu representations"; see Ruth Millikan, "Pushmi-Pullyu Representations," in *Philosophical Perspectives 9: AI, Connectionism, and Philosophical Psychology*, ed. James E. Tomberlin (Atascadero, Calif.: Ridgeview Publishing, 1995).

[28]Churchland, Ramachandran, and Sejnowski, "A Critique of Pure Vision," 60.

[29]J. DiPelligrino, R. Klatzky, and B. McCloskey, "Time Course of Preshaping for Functional Responses to Objects," *Journal of Motor Behavior* 21 (1992): 307–316.

[30]See G. Rizzolatti, L. Fadiga, and L. Fogassi, "Premotor Cortex and the Recognition of Motor Actions," *Cognitive Brain Research* 3 (1996): 131–141.

[31]Marc Jeannerod, *The Cognitive Neuroscience of Action*, 191.

[32]Donald Michie and Rory Johnson, *The Creative Computer: Machine Intelligence and Human Knowledge* (London: Penguin Books, 1984), 95.

[33]See Pylyshyn, *The Robot's Dilemma*, for more discussion.

[34]Dana H. Ballard et al., "Deictic Codes for the Embodiment of Cognition," *Behavioral and Brain Sciences* (forthcoming).

[35]Ibid., 6.

[36]This example is borrowed from Andy Clark, "I am John's Brain," *Journal of Consciousness Studies* 2 (2) (1995): 144–148.

[37]The story is detailed in Michael Triantafyllou and George Triantafyllou, "An Efficient Swimming Machine," *Scientific American* 272 (3) (1995): 64–71, and further discussed in Clark, *Being There*. A 49-inch, eight-segment anodized aluminum and lycra robot tuna is being used at MIT to test the details of the theory.

[38]For more on cognitive extension, see Daniel C. Dennett, *Darwin's Dangerous Idea* (New York: Simon and Schuster, 1995), chaps. 12 and 13.

[39]Dorothy Edwards, Carolyn Baum, and N. Morrow-Howell, "Home Environments of Inner-City Elderly: Do They Facilitate or Inhibit Function?" *The Gerontologist: Program Abstracts of the Forty-Seventh Annual Scientific Meeting* 34 (1) (1994): 64; Carolyn Baum, "Addressing the Needs of the Cognitively Impaired Elderly from a Family Policy Perspective," *American Journal of Occupational Therapy* 45 (7) (1991): 594–606.

⁴⁰Some of this additional structure is maintained and provided by family and friends. (But, similarly, much of our own wideware is provided by language, culture, and institutions that we do not ourselves create.)

⁴¹See Clark, *Being There,* chart 10. Such counterbalancing is, as Marcel Kinsbourne has usefully reminded me, a somewhat delicate and complex matter. The mere provision of the various props and aids is useless unless the patient remains located in a stable, familiar environment. And the ability of different patients to make use of such added environmental structure itself varies according to the nature and extent of the neurally based deficit.

⁴²See Dawkins, *The Extended Phenotype.*

⁴³See W. Kutsch et al., "Wireless Transmission of Muscle Potentials During Free Flight of a Locust," *Journal of Experimental Biology* 185 (1993): 367–373. My thanks to Joe Faith for drawing this example to my attention.

⁴⁴See, for example, Esther Thelen and Linda B. Smith, *A Dynamic Systems Approach to the Development of Cognition and Action* (Cambridge, Mass.: MIT Press, 1994), Robert F. Port and Timothy Van Gelder, *Mind as Motion: Explorations in the Dynamics of Cognition* (Cambridge, Mass.: MIT Press/ Bradford Books, 1995).

⁴⁵A move suggested by Edwin Hutchins, *Cognition in the Wild* (Cambridge, Mass.: MIT Press, 1995).

⁴⁶See Clark, *Being There;* Jim Crutchfield and Melanie Mitchell, "The Evolution of Emergent Computation," *Proceedings of the National Academy of Sciences* 92 (1995): 10742–10746; M. Mitchell, J. Crutchfield, and P. Hraber, "Evolving Cellular Automata to Perform Computations," *Physica D 75* (1994): 361–391; and Timothy Van Gelder and Robert F. Port, "It's About Time: An Overview of the Dynamical Approach to Cognition," in *Mind as Motion,* ed. Robert F. Port and Timothy Van Gelder (Cambridge, Mass.: MIT Press, 1995), 1–44.

⁴⁷Jerry A. Fodor, *The Language of Thought* (New York: Crowell, 1975). See also Jerry Fodor and Zenon Pylyshyn, "Connectionism and Cognitive Architecture: A Critical Analysis," *Cognition* 28 (1988): 3–71.

⁴⁸I am very grateful to Stephen Graubard and the participants at the *Dædalus* authors' meeting in Paris in October 1997 for a wealth of useful advice, good criticism, and wise counsel. Special thanks to Jean-Pierre Changeux, Marcel Kinsbourne, Vernon Mountcastle, Guilio Tonini, Steven Quartz, and Semir Zeki. As usual, any remaining errors are all my own.

Subject Index

Illustrations

The Brain

Plate 2   *Illustrations*

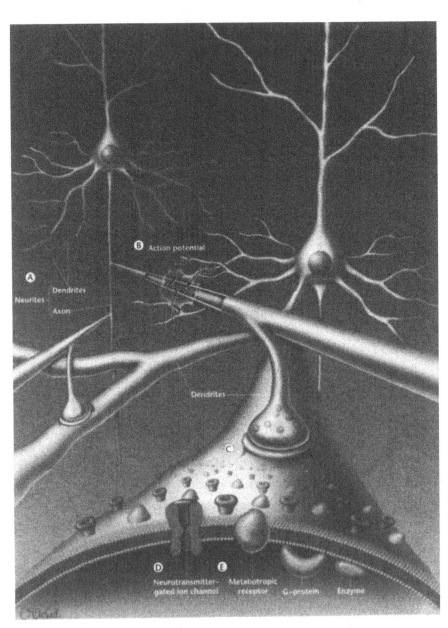

Figure 1: Membranes of the mind. The information-processing cells of the brain—the neurons— are enveloped in a thin sheet of phospholipid membrane, 5nm thick. The brain conducts most of its business along these surfaces, particularly along membrane-e~lclosed tubes called neurites (A). There are two types of neurite, axons and dendrites. Axons are specialized to function as the wires along which information is conducted from one place to another in the nervous system. Dendrites are specialized to detect chemicals released by axons at sites of axon-dendrite contact, called synapses. These functional specializations of axons and dendrites arise from the different types of protein molecules found embedded in their membranes.

The membrane proteins of axons enable them to conduct nerve impulses, or action potentials (B). When a neuron is not generating nerve impulse— and is therefore at rest—the inside of the membrane has a slight negative electrical charge compared to the outside of the membrane. The action potential is a brief reversal of this situation, so that for a millisecond or so the inside of the membrane becomes positively charged with respect to the outside. This reversal depends on membrane proteins called voltage-gated sodium channels. As the name suggests, these proteins form channels or pores in the axonal membrane through which only sodium ions can pass. The channels are gated— opened and closed—in response to a change in the electrical charge difference (voltage) across the membrane. Sodium is a positively charged ion and is in great abundance outside the neuron. Thus, when the channel opens, sodium enters the axon, and the inside of the membrane assumes a positive charge rapidly (the change in membrane voltage is depolarization). The resulting action potential is brief mainly because the channels are only open for an instant. Viewed from this perspective, the action potential is a brief intracellular sip of sodium ions from a salty extracellular sea.

The sodium wave that is the action potential propagates along the axon. Propagation occurs because even as the channels are closing at one position in the membrane, they are being stimulated by the positive charge to open in the membrane just ahead. The impulse sweeps without decrement from its origin, usually back in the cell body, all the way to the axon terminal, which can be more than two meters away. Because all action potentials are the same duration and the same "size" (that is, they consist of the same voltage change), a frequency code is employed to transfer information, much as in FM radio. The highest "firing frequencies" can approach 1,000 impulses per second. Action potentials travel at different rates, depending on the properties of the axon, but the fastest call exceed 100 meters per second.

Often along the course of the axon, and always at its end, are synapses—the specialized sites of interneuronal contact (C). Here the depolarization of the membrane stimulates the release of chemical neurotransmitters. The neurotransmitters are stored in spherical bags of membrane called synaptic vesicles. In response to the action potential, the vesicle membrane fuses with the surface membrane of the axon, thus releasing the vesicular contents into the extracellular space.

Neurotransmitter release requires the entry of calcium ions into the axon. Entry of calcium is controlled by voltage-gated calcium channels in the axon membrane that function much like their cousins, the voltage-gated sodium channels. Unlike sodium, however, calcium is more than an electrical-charge carrier. Calcium ions can stimulate a wide variety of biochemical reactions by binding directly to various proteins. Calciumbinding proteins exist within the membrane of the synaptic vesicles, and these are believed to serve as the triggers for vesicular fusion and neurotransmitter release. There are many different chemical neurotransmitters, ranging from small amino acids and amines to large peptides. All act at the postsynaptic, usually dendritic, membrane by binding proteins called neurotransmitter receptors.

There are an enormous number of different receptors, but they can be divided into two broad categories. One type is the neurotransmitter-gated ion channel (D). These channels open only when bound to a transmitter. Depending on the charge of the ions that pass through the open channel, and the direction of net charge movement, the neurotransmitter can excite (depolarize) the postsynaptic membrane or inhibit it (keep it at a negative voltage). The second type of receptor is called a metabotropic or G-protein-coupled receptor (E). These receptors initiate cascades of intracellular biochemical reactions that can have diverse effects on the postsynaptic neuron, including the stimulation of gene expression and protein synthesis. Most neurons have thousands of synapses that excite, inhibit, or permanently transform it. All these inputs are integrated by the dendrites, and the voltage changes of the dendritic membrane are translated into action potentials by the axonal membrane.

A typical neuron with all its neurites has a membrane surface area of about 250,000 mm2.* The surface area of the 28 billion neurons that make up the human cerebral cortex comes to 7,000 m2—roughly the size of a soccer field. This expanse of membrane, with its myriad specialized protein molecules, constitutes the fabric of our minds.

S. Cullheim, J. W. Fleshman, L. L. Glenn, and R. E. Burke, "Three-Dimensional Architecture of Dendritic Trees in Type-Identified a-motoneurons," Journal of Comparative Neurology 255 (1987): 8296.

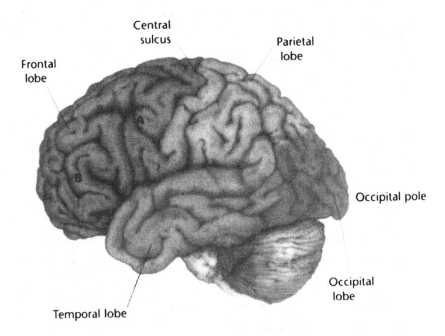

Plate 4   *Illustrations*

Central
sulcus

Parietal
lobe

Frontal
lobe

Occipital pole

Occipital
lobe

Temporal lobe

Figures 2 and 3: The anatomy of the brain. The two diagrams here and on the following page, present the brain viewed a) from the side and b) from the inside. The image above is of a preserved specimen; that on the following page presents a view obtained by splitting the brain lengthwise down the middle and viewing the inside, cut surface. The cut can be made postmortem with a real knife or, as in this picture, in vivo with a virtual knife, the computer. (Note the nose to the left of the picture indicating the front of the head.) The main lobes are indi-

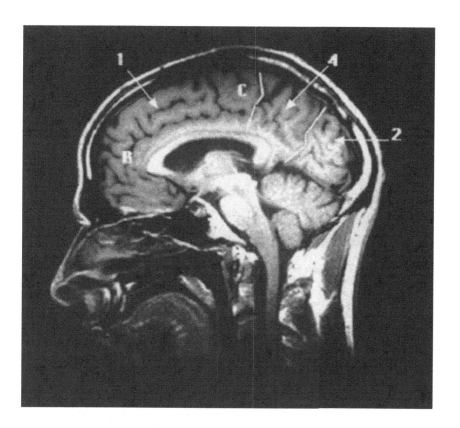

cated on both the side and inside views, as follows: frontal (1); occipital
(2); temporal (not shown in figure 3); parietal (4). The most posterior
part of the occipital lobe is known as the occipital pole. In addition,
some of the landmarks associated with the motor system are
indicated: (A) premotor cortex; (B) prefrontal cortex; and (C)
supplementary motor area.

Plate 6   *Illustrations*

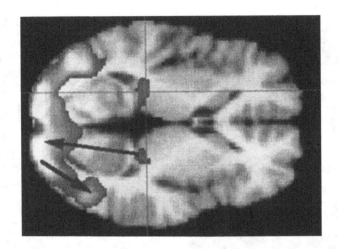

Figure 4: The insights provided by functional MRI. An image of a slice of the brain is shown. The front of the brain is shown pointing to the right. The left side is uppermost. This is a thin slice scanned with MRI to show the brain anatomy It is roughly at a level that joins the occiput with the bridge of the nose and the ear canals. Superimposed are areas shown to be active with functional MRI (fMRI). They have been defined by a comparison of activity when moving and stationary objects were viewed. The top arrow starts in a deep small structure that is also present on the other side (and through which the crosshairs pass), known as the lateral geniculate nucleus. It is the main relay station for nerve fibers carrying signals from the retina of the eye to the occipital cortex. The arrow points to a very active area at the back of the brain, which is the primary visual cortex (V1). This part of area V1 carries a map of the center of the field of view— that part of vision with the highest acuity that we use to fix objects and to examine them in detail. Signals then pass through area V2 to another more lateral blob of differential activity that is also present on both sides of the brain. This is area V5 in which activity is associated with the perception of visual motion. Though the arrows suggest one-way traffic, functional imaging has shown that there are active reciprocal pathways whose strength can be measured.

Encoding

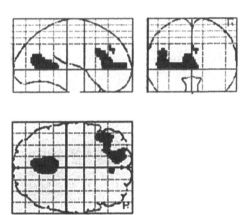

Figure 5: Three sets of three functional maps of the brain. The three examples are given in a common format: the top right image is a view through a transparent standard brain from the back with the top of the brain uppermost. Next to it, on the left, is the equivalent view from the side with the front of the brain pointing right. Below it is the view from above with the left side of the brain uppermost. This format allows one to look at the three-dimensional distribution of differentially activated areas, shown as black blobs.

The standard brain is divided up by a coordinate grid that allows one to localize a given area of activation. These maps are known as statistical parametric maps (SPM) in the jargon of brain cartographers. The set above, labeled "Encoding," shows areas of the brain associated with the acquisition of autobiographical memories. Of note is the area in the frontal lobe on the left and a second area at the back of the brain that lies in the midline in a region known as the retrosplenial cortex. It is of interest because patients with tumors localized to this region have memory disturbances. On the next page, the two

Plate 8   *Illustrations*

Retrieval

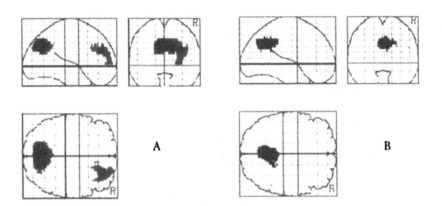

A

B

sets labeled "Retrieval" are both associated with the retrieval of autobiographic memories. On the left (A) is a pattern of right frontal activation, associated in this case with a posterior area that is situated higher in the brain, in the midline parietal cortex. The function of this area has been partially dissected with a second experiment in which the differential activation associated with recalling easily imageable words (such as the names of objects) was contrasted with the activity related to the recall of words that are difficult to associate with mental images (for example, content and abstract words such as "faith"). It can be seen on the right (B) that the parietal activity (in a region known as the precuneus) is associated specifically with remembering easily imageable words. In addition, there is no differential activation of the right frontal cortex, indicating that it is equally active during the recall of either type of word.

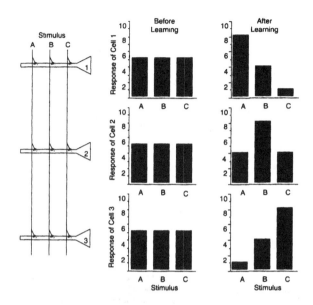

Figure 6: A model of distributed information storage. Three neurons (1, 2, 3) receive inputs carrying information about 3 stimuli (A, B. C). Before learning, all neurons respond equally to all stimuli. After learning, the neurons show stimulus selectivity, reflecting the modification of synapses in the network. (Reproduced with permission from Mark Bear, "A Synaptic Basis for Memory Storage in the Cerebral Cortex," *Proceedings* of *the National Academy* of *Sciences* 93(1996):13453-13459).

Presynaptic input activity: **d**

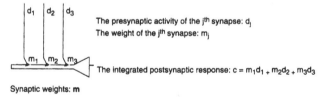

The presynaptic activity of the jth synapse: d_j

The weight of the jth synapse: m_j

The integrated postsynaptic response: $c = m_1 d_1 + m_2 d_2 + m_3 d_3$

Synaptic weights: **m**

Change in the weight of the jth synapse: $\dot{m}_j = \phi(c)\, d_j$

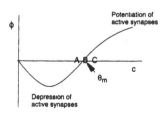

Figure 7: The BCM function controlling synaptic plasticity at cortical synapses.

Plate 10   *Illustrations*

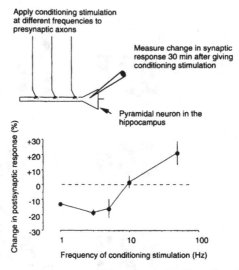

Figure 8: Effects of conditioning stimulation delivered to the Schaffer collaterals in the CA1 region of hippocampus at different frequencies. Plotted is the mean (±SEM) effect of 900 pulses of conditioning stimulation delivered at various frequencies on the response measured 30 minutes after conditioning. (Modified from Serena Dudek and Mark E Bear, "Homosynaptic Long-Term Depression in Area CA1 of the Hippocampus," *Proceedings of the National Academy of Science* 89 (1992): 4363-4367.)

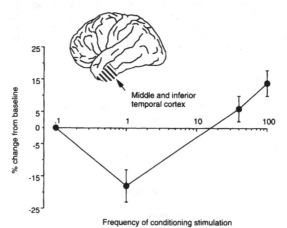

Figure 9: Bidirectional plasticity of synaptic responses in adult human inferior temporal cortex. The frequency-response function was derived in a manner similar to that shown in figure 8, using slices of human inferior temporal cortex resected during surgery. These data were replotted from Wei Chen et al., "Long-Term Modifications of Synaptic Plasticity in the Human Inferior and Middle Temporal Cortex," *Proceedings of the National Academy of Science* 93 (1996): 8011-8015.

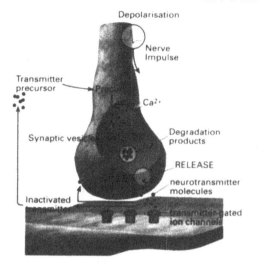

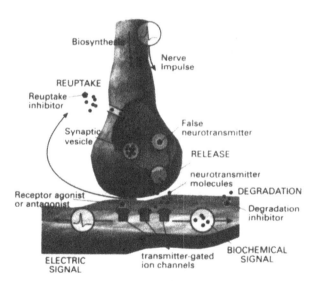

J. Christian Hoffman

Figure 10: Chemical transmission of signals in the brain and principal targets for drugs of abuse. (Compare with the process of synaptic transmission described in Figure 1.)

Plate 12   *Illustrations*

Morphine

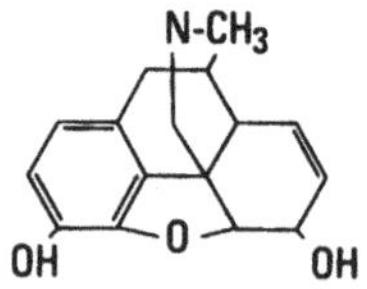

Anandamide

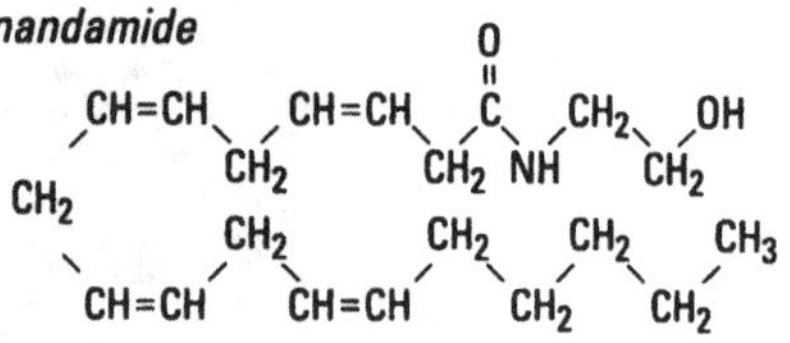

Leu-Enkephalin

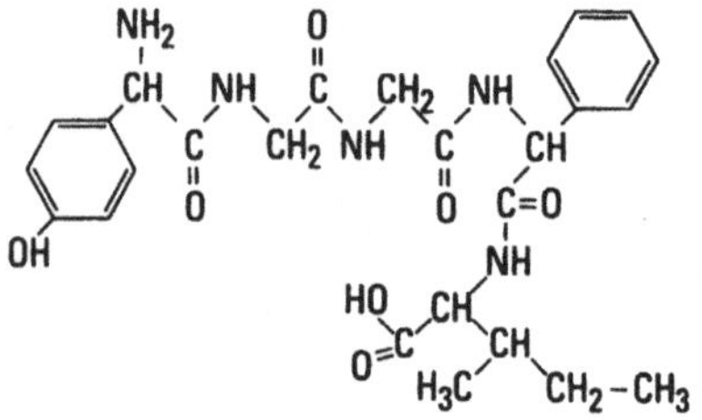

Δ 9 THC

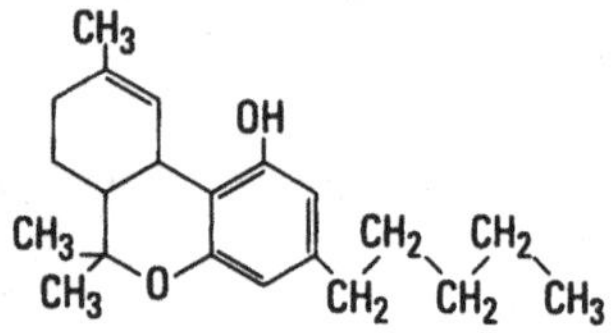

Figure 11: Structural analysis comparing neurotransmitters and drugs of abuse.

Next page:
Neurotransmitter receptors as targets of drugs of abuse.
Figure 12 (top): Channel-linked receptor, such as nicotine receptor.
Figure 13 (bottom:) G-protein-linked receptor, such as opiate receptor.

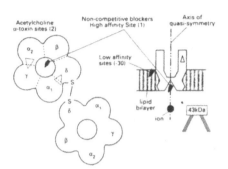

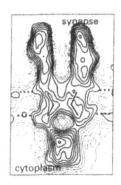

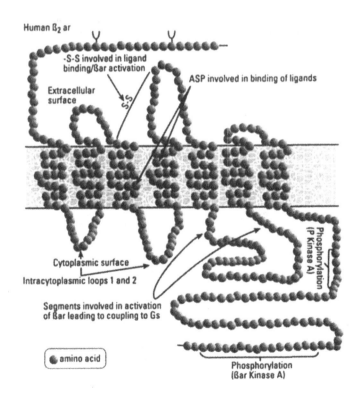

Plate 14   *Illustrations*

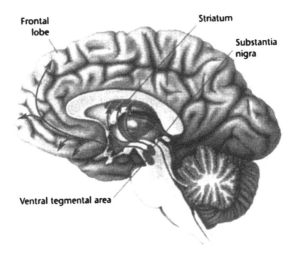

Frontal
lobe

Striatum

Substantia
nigra

Ventral tegmental area

Figure 14: Reward circuits in the brain: the dopaminergic diffuse modulatory systems arising from the substantia nigra and the ventral segmental area. The substantia nigra and the ventral segmental area lie close together in the midbrain. They project to the striatum (caudate nucleus and putamen) and limbic and frontal cortical regions, respectively.

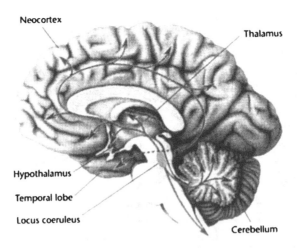

Neocortex

Thalamus

Hypothalamus

Temporal lobe

Locus coeruleus

Cerebellum

Figure 15: The norepinephrine system. Shown is the noradrenergic diffuse modulatory system arising from the locus coeruleus. The small cluster of locus coeruleus neurons project axons that innervate vast areas of the CNS, including the spinal cord, cerebellum, thalamus, and cerebral cortex.

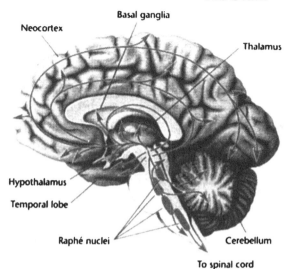

Figure 16: The serotonergic diffuse modulatory systems arising from the rapine nuclei. The rapine nuclei are clustered along the midline of the brain stem and project extensively to all levels of the central nervous system.

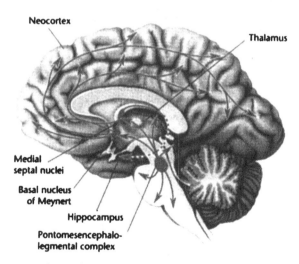

Figure 17: The acetylcholine system. Shown is the cholinergic diffuse modulatory systems arising from the basal forebrain and brainstem. The medial septal nuclei and basal nucleus of Meynert project widely upon the cerebral cortex, including the hippocampus. The pontomesencephalotegmental complex projects to the thalmus and parts of the forebrain.

Plate 16   *Illustrations*

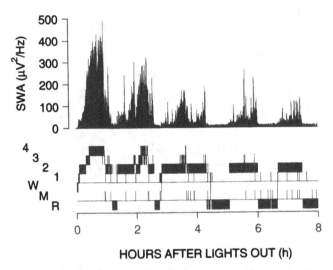

Figure 18: Sleep Pressure. Habitual short sleepers live at a higher level of "sleep pressure" than long sleepers and respond less to sleep deprivation. The time course of SWA (mean values with standard errors) and simulated Process S (interrupted curves) is show for two baseline wake-sleep cycles and a sleep deprivation period followed by recovery sleep. As predicted by the model, short sleepers showed a smaller increase of SWA after the extended waking period than long sleepers. Modified from Borbély and Tononi, note 27.

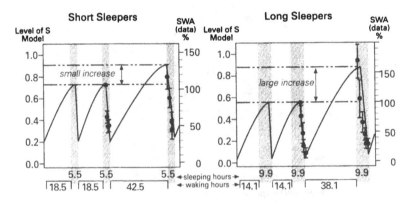

Figure 19: Sleep as a dynamic process. The dynamics of sleep are here reflected by slow-wave activity (SWA; power density in the 0.75-4.5 Hz range) in the electroencephalogram (EEG), which is a measure of non-REM sleep intensity. It consists of five non-REM/REM sleep cycles. The sleep profile below the SWA plot depicts stages 1-4 of non-REM sleep, waking (W), movement time (M), and REM sleep (R).

Printed in the United States
by Baker & Taylor Publisher Services